W9-COW-726

THE
COSMIC
MACHINE

THE
COSMIC
MACHINE

THE SCIENCE THAT
RUNS OUR UNIVERSE
AND THE STORY
BEHIND IT

SCOTT BEMBENEK

San Diego

Copyright © 2017 Scott Bembenek

All rights reserved. This book was professionally published by the author Scott Bembenek through Zoari Press. This book or parts thereof may not be reproduced in any form, stored in any retrieval system, or transmitted in any form by any means – electronic, mechanical, photocopy, recording, or otherwise – without prior written permission of the author. If you would like to seek written permission, or have other questions about the book, contact the author at:

scottbembenek.com

The Cosmic Machine:
The Science That Runs Our Universe and the Story Behind It

Published by
Zoari Press
San Diego, CA, USA

Line Editing: BubbleCow (*bubblecow.com*)
Copy Editing & Proofreading: Sabrina Leroe (*bespoke-editorial.com*)
Interior Design: Wordzworth Ltd. (*wordzworth.com*)
Cover Design: Derek Murphy (*creativindiecovers.com*)
Figures: Kanwar Singh (*fiverr.com/kanward*)
Index: Jen Weers (*jenweers.com*)
Author Photo: Sherry Kostecka (*sherrykosteckaphotography.com*)

ISBN: 978-0-9979341-0-6 (paperback)
        978-0-9979341-3-7 (hardcover)
Library of Congress Control Number: 2016913451

First Edition
Printed in the United States of America

For Ariana and Zoey,

May you find the answers to *all* of life's difficult questions.

About the Author

Scott Damian Bembenek grew up in the small town of Delavan, Wisconsin (population of approximately 5,000 at that time). Scott's passion for science began around the age of five. Throughout his childhood, "science" took on a variety of forms, involving such things as batteries, circuits, magnets, anything electronic, insects, frogs, toads, salamanders, turtles, fish, etc. He conducted new "experiments" on a weekly (sometimes daily) basis, much to the astonishment of his (exceptionally patient) mother.

Scott attended Carroll College (now Carroll University) in Waukesha, Wisconsin. There he further pursued his interest in chemistry, especially organic chemistry. However, by the end of his third year it was clear that his real interest was physical chemistry, a more theoretical, less laboratory-intensive version of chemistry. Indeed, Scott's skills weren't in the laboratory, which should have been clear to him much sooner since the laboratory portion of classes always brought down his grade – unlike for his fellow students. During this time his interest in physics grew. In his last year in college, Scott was fortunate to be able to take graduate-level classes and conduct research in physics under the direction of his advisor, as part of a self-designed curriculum toward a BS in chemistry and physics, which he received in 1993.

Scott did his graduate work at the University of Kansas in Lawrence, where he investigated the dynamics of complex liquid systems using computer simulations and theoretical techniques. His work on supercooled liquids and glasses remains well cited even today. Scott received his PhD with honors in 1997 in theoretical chemical physics.

Thereafter, he was awarded a National Research Council Fellowship and joined the Army Research Laboratory in Aberdeen, MD. There he used computational and theoretical techniques to study the dynamics of explosives. Scott pioneered a new approach to parameterizing interaction potentials necessary for accurate computer simulations. His work was well recognized, and upon the completion of his fellowship, he was immediately offered a full-time position. However, Scott had other plans.

Intent on pursuing a career in academia, Scott took a research associate position (and a 50% salary cut) at Colorado State University. While computer simulations once again played a role in his work, the majority of his time at the office involved "old-school, paper-and-pencil" theoretical physics on fluid systems. In fact, his paper *A Kinetic Theory for Dilute Dipolar Systems* required more than 300 pages of derivations! When his term at Colorado State was finishing up after two years, he began looking for permanent employment. He attended a recruiting session at the university held by Johnson & Johnson that gave an overview of the drug discovery research at its San Diego site and potential employment opportunities. Although scheduled for an interview the following day, he had intended to pass on the opportunity. Some wise advice from a friend convinced him otherwise.

Currently, he is a Principal Scientist in the Computer-Aided Drug Discovery group at Johnson & Johnson Pharmaceutical Research & Development in San Diego, CA, where he has worked since 2002. During his time there, he has made substantial contributions to numerous drug discovery projects and has worked in a variety of disease areas.

scottbembenek.com

Contents

Preface

As I sat with the "final version" of the manuscript, looking it over for "one last time" (a lie I had told myself so many times before, and each time fully believing it), I sighed; this time I had it. A brief conversation with a friend had started me down this long, often wayward, journey of writing my first popular science book, or, as I like to call it, a "science story." My friend had recounted to me something they had read by a popular science writer. I was intrigued – not so much by the story itself (honestly, I have forgotten both the writer and the story), but rather by my friend's excitement with what they had read. Let's face it: science doesn't often incite this response in the majority of people.

So, as a scientist myself, it was uniquely rewarding for me to witness that enlightened moment my friend had experienced; I wanted to create those types of insightful science moments for others. Moreover, there's a long, rich history of pivotal scientific achievements that have been clouded or simply ignored by the current popular science literature, and I wanted to bring these to light as well.

It was these reflections that motivated me to write a popular science book. Thus began a journey that started in August of 2009. I was originally convinced that I could finish this project in two years. However, in the ensuing two years many things happened: I suffered paralysis in my left hand in a martial arts training accident but continued to write, before surgery and during the almost two-year recovery process, albeit a bit at a slower pace; got married and had four ceremonies (two in San Diego, one in Wisconsin, and one in China); took on more responsibility at work (I still have my day job); lost my mom; and became a dad. While all these things undoubtedly contributed to the extended time frame, they're not the main reasons.

I had a particular vision for the book, and I refused to publish it until it was realized. This, in addition to my bit of OCD and touch of perfectionism, resulted in a much-extended timeline. Nonetheless, I think (and I hope you do too) it was worth it.

As I see it, a major stumbling block for the popularity of science is its presentation. It's often presented in a very dry, dull manner. Many of you have taken *those* classes, seen *those* videos, heard *those* talks … and so have I. Clearly, for me (and other scientists) those things weren't major setbacks – we became scientists, after all. Sure, we found some of those things boring, too (yes, scientists are human beings), while other parts of it we found quite exciting. For me the bottom line is this: science is a passion.

My passion for science began when I was very young, around five years old. Back then "science" was about taking apart electronics and performing experiments – all humane! – on insects and garden-variety suburban amphibians. In fact, I will never forget when in second grade I went to our (very small) school library to check out a book on how electricity worked. In a quite discouraging tone the librarian said, "Wouldn't you much rather check out a book with a nice story?" To which I replied, "Nope."

For most of my childhood I suffered from pretty bad allergies and asthma, which meant I missed a lot of school. The routine that was set up early in my academic life was this: I stayed home from school, Mom picked up my homework from my teachers, and I worked on it at home – sort of a power version of homeschooling, if you will. And I always worked ahead when it came to my science and math classes. I guess you can say that science kept me engaged in school during those rather difficult years, providing me with a perspective that persisted into adulthood and has remained with me ever since.

You see, I've been a scientist for a long time now, officially since 1997 when I got my PhD in theoretical chemical physics. Like I said, I became a scientist because I have a passion for it. I have a passion for the way it allows you to see the world, the way it gets you thinking about the world, and the way it allows you to understand the world. I also believe that this unique perspective is open to everyone, scientist or not.

My goal for this book is simple: to make the major aspects of physics and chemistry more accessible to nonscientists, and in doing so, to spark an interest in science – or even better, an excitement for it. It's often said that science isn't for everyone. Sure, not everyone will have *my* passion for science, but I still think that science has something to offer to just about everyone. In short, this book is my best effort to help you find at least a tiny bit of wonder in science.

Acknowledgments

Book writing is not an "adventure" one pursues alone. The seemingly wayward path becomes more defined through the support of friends and family; it's these people I'd like to acknowledge.

I'd like to thank my mother, who endured countless "science experiments" (a.k.a. "the project of the week") throughout my formative years. These inquiries, as haphazard as they often were, provided me the growing room I needed early on to explore my love of science.

Then there's my darling Sophia. I had only just begun this book-writing journey (even though I thought I was wrapping things up) when we met. She supported me every step of the way and provided the often-needed nudge to see this book through to the end. I am forever grateful.

There were several people who read through various versions of the manuscript and provided insightful feedback: Zachary Bachman, Elizabeth Nguyen, Ruthi Swartzberg, Nicholas Evergates, Frank Axe, Jess Behrens, and Scott P. Brown. I'd like to give special thanks to Erik Shipton for taking the extra time to critically review the scientific aspects of the manuscript.

Finally, there were those who offered much-appreciated encouragement over the years, in particular, Jim Dahl, Robb Hankins, Marty Jenkins, and KC Chea.

Introduction

This book is not a textbook and therefore is not laid out as one. There are no problems to work through at the end of a chapter, and no rigorous derivations of any sort are shown. I leave those to the excellent references provided in the bibliography. The book is written as a "science story." That is to say, I use the history around the scientific discoveries and the biographies of scientists to shape an interesting story as I weave in the actual science throughout its course. This will allow you to become familiar with the "key players" in the history of science, what they struggled with at the time (personally and scientifically), and the way we understand natural phenomena today.

The book is divided generally into four major topics:

- Energy
- Entropy
- Atoms
- Quantum Mechanics

I chose these because I consider them some of the most (if not the most) important topics in physics and chemistry. If you understand these areas, you've got a good handle on a lot of science. They also happen to be some of the most interesting, in my humble opinion.

These four key topics are broad, and so I further divide them into chapters and sections to make the material very approachable. You don't need to start with the first topic; you can start with any one you like. However, I would recommend completing one topic before moving on to another. The point is that each is written to be (mostly) self-contained while at the same time being well connected to the other topics covered in the book. In this way, you can both learn about a given topic in detail and further your knowledge of it by understanding its connection to the other topics. For those of you who just have to know the "formal areas" of physics and chemistry that the book touches upon, they are classical mechanics, thermodynamics, statistical mechanics, kinetic theory, quantum mechanics, and many more.

You'll notice that I include many footnotes throughout the book. As much as I do cover, I always find myself wanting to discuss some topics more. In these cases, I decided the extra discussion would better serve the book as a footnote rather than the main text. The information in the footnotes not only gives a broader perspective of a discussion, it also provides more detail. In some cases, this material is more advanced, building on the main text, which will give a little more satisfaction for the science enthusiasts and experienced scientists.

OK, let's talk about the elephant in the room, or the book, rather. Although there are no derivations, I still show a few equations here and there, albeit very sparingly, primarily in the later chapters. To be honest, I struggled with putting any equations in at all. Indeed, Stephen Hawking was warned by an editor that for every equation he included in *A Brief History of Time* his readership would be halved; in the end, Hawking included only a single equation. To be perfectly honest, I wrote the text so that it stands on its own without the equations. Why then did I include those equations? My motivation for doing so was to illustrate how mathematics and science complement each other – together they often provide the best overall explanation of physical phenomena. As Richard Feynman so eloquently puts it in *The Character of Physical Law*:

> If you want to learn about nature, to appreciate nature, it is necessary to understand the language [of mathematics] that she speaks in. She offers her information only in one form; we are not so unhumble as to demand that she change before we pay any attention.

Nonetheless, I want you to feel free to **skip over any, even all, of the equations** – you're not going to miss much if you do – rather than being put off by their presence.

So, if you're looking for an interesting story, with some history, biography, and science, this is the book for you. Even if you're not already science-savvy, you'll be able to gain insight into several areas of physics and chemistry. And if you're already somewhat of a scientist, this book will help you find a new appreciation and perspective for topics already familiar to you, and you'll probably learn a few things you didn't already know (especially if you avail yourself of the extra information in the footnotes). If nothing else, it's a good "science story" that most people will enjoy.

Scott Bembenek (Fall 2016)

THE
COSMIC
MACHINE

PART I

The First Law: Energy

It is important to realize that in physics today,
*we have no knowledge of what energy **is** ....*
It is an abstract thing in that it does not tell us
*the mechanism or the **reasons** for the various formulas.*

– RICHARD FEYNMAN, AMERICAN PHYSICIST (1918–1988)

CHAPTER 1

Nothing for Free

The Conservation of Work

At some basic level we all understand energy. We're aware of the various forms of energy that influence our lives: the gas needed to drive our cars; the electricity to run our TVs, toasters, refrigerators, stoves, and other appliances; the batteries in our cameras, remotes, cell phones; and the list goes on and on. Simply put, energy is a fundamental physical property of every *system*. In this way, it's much like other physical properties we use to describe a system, such as temperature, pressure, and density. As for the system, it can be anything: someone driving a car down the highway, a hot cup of coffee, or the Pacific Ocean. We can meaningfully talk about, and sometimes even determine, the energy of a system.

Energy itself is very elusive because it takes on many different forms and can seamlessly transform itself between these forms without clearly revealing the slightest hint of ever doing so. As physical properties go, energy truly is a chameleon. This characteristic behavior of energy also comes through in the subtle way we talk about it: its ability to do a certain type of work, create motion, and change the temperature. Its vague nature, our challenges in describing it, and the lack of a clear picture of the composition of matter are the reasons that recognizing and understanding energy was such a substantial undertaking.

So substantial was this task that in addition to not understanding energy for so long, we couldn't even agree to call it "energy" until about 1850. Much has changed since then, and today our understanding of energy is quite impressive. Just look at all the various machines we've created that make our lives easier and improve our overall quality of life; modern technology is truly remarkable. Although the first machines ran solely on the labor of human beings or animals, these *simple machines* not only made life easier, they also provided the very first insight into the way energy behaves.

Nature's Compensation and Simple Machines

For ancient people, hard manual labor was an inevitable part of life. Embracing this harsh reality, they got clever and developed simple machines (lever, inclined plane, screw, pulley, wheel, and wedge) to give a helping hand in their necessary endeavors. These devices must have seemed magical back then, as objects that were once only moved with a great effort (or not at all) were now easily moved when using a simple machine. However, one enduring fact was painfully clear: the *mechanical advantage* afforded by using these devices always came with a price, which can be considered as compensation or payment to the universe for the job being completed with *less effort* on one's part.

Let's get a better understanding of simple machines by considering the inclined plane, which is simply a ramp (like a wheelchair ramp) used to move an object up to a desired height. Its sole purpose is to make reaching that height easier than if one were to simply move to it from directly below. The inclined plane has benefited many civilizations throughout history. Prehistoric people used the inclined plane to help them move heavy objects. The construction of the Egyptian pyramids benefited in a similar fashion by using the inclined plane. The ancient siege ramp, an essentially military weapon, allowed its attacker to more easily traverse the wall of his enemy's fortress. Unlike with the other simple machines, the use of the inclined plane doesn't actually involve any motion on its part. In other words, an applied force to the inclined plane itself isn't required to "implement" it. Therefore, the mechanical advantage afforded by the inclined plane is simply due to one walking the distance of its slope. This lack of necessary motion is most likely why the inclined plane wasn't recognized as a simple machine – it was the last of the six to be recognized – until several Renaissance mathematicians had worked out its mechanical advantage.

Anyone who has spent time climbing stairs is already familiar with the inclined plane; stairs are just a fancier version. By using the stairs you're able to move from a lower to higher height with *less effort*, or, more specifically, by using *less force*. That's all there's to it. It's the same thing when using an incline plane to move an object from a lower height. Now that less force is required, objects that were impossible to budge are moved with ease, while those that were somewhat difficult are very easily moved. But there's a catch: you will have to move the object farther than before. That is to say, if you want to use the inclined plane to help you move an object (and who wouldn't?), then you have to move the object over a longer distance to get to the desired height than if you had started from directly below and moved upward. This is probably already clear to you from a lifetime of stair climbing.

Consider all the stairs you climb compared to the actual height you reach from where you started. This height is **always** less than the distance you climbed in stairs. In other words, more distance in stairs is traded for less force to reach the intended height. Now, if we were to pass on the stairs altogether and simply climb straight up to your destination (from directly below it), it would be a shorter climb for sure, but the needed force to do so would be greater. Therefore, we have stairs in our homes rather than ladders.

In this way, the inclined plane is not unique: as it turns out, less required force on one's part being traded for movement over a greater distance is the common theme among **all** six of the simple machines.

Force, Distance, and Work

From our previous discussion, it appears that we've come across a relationship between the required force and the distance traversed when attempting to rise up to a certain height via an inclined plane. Let's state it very clearly:

the force required to move an object up an inclined plane (or stairs) is less than the force needed to move the *same* object up a ladder to the same height. In other (more mathematical) words:

$$F_{\text{inclined plane}} < F_{\text{ladder}}$$

where F is the force and < means "less than." Further, the price we pay for this luxury of exerting less force is an increase in the distance we must travel. That is,

$$d_{\text{inclined plane}} > d_{\text{ladder}}$$

where > means "greater than." In our discussion, the object being moved is simply you, but in general it could be anything; perhaps we're carrying or pushing something. Regardless, the same relationship between the force and the distance **always** holds.

The *inequalities* above give us nice, compact relationships between force and distance. From them we more easily see that as one "goes up," the other "goes down." That is to say, there's a sort of compensatory effect between the force and the distance. In fact, it turns out that these effects are perfectly balanced, such that regardless of if we use an inclined plane or a ladder to move an object to a certain height, it takes the same amount of *work*:

$$Work = \left(force\ moving\ object \right) \times \left(distance\ object\ is\ moved \right)$$

Therefore, in terms of the work involved in using an inclined plane versus that in using a ladder:

$$Work_{\text{inclined plane}} = Work_{\text{ladder}}$$

This means the work required to move something to a given height is *conserved*. In other words, nature doesn't really care how you get it there; the amount of work required is still the same, no more, no less.

This becomes clearer when we take a moment to consider exactly why it takes work in the first place to move something to a higher height. What are we working against? We're working against the gravitational force of the Earth,[1] and raising something to a higher height increases its *potential energy*. We'll discuss potential energy in much more detail later, but for now we note that work and energy have a close relationship. Moreover, as such we also begin to suspect that nature has a tendency to conserve energy as well.

It's tempting to imagine that we might be able to create a machine that would allow us to use less force to move an object without having to move the extra distance required. Unfortunately, there's no "free lunch" when it comes to the universe, and this machine will never exist in reality. Perhaps no one stated this notion clearer than Galileo Galilei (1564–1642):

> I have seen (unless I am much mistaken) the general run of mechanicians deceived in trying to apply machines to many operations impossible by their nature, with the result that they have remained in error while others have likewise been defrauded of the hope conceived from their promises. These deceptions appear to me to have their principal cause in the belief which these craftsmen have, and continue to hold, in being able to raise very great weights with a small force, as if with their machines they could cheat nature, whose instinct – nay, whose most firm constitution – is that no resistance may be overcome by a force that is not more powerful than it.

Nonetheless, many have tried (and still try) to "cheat" the universe in one way or another. A particular example of this is the "perpetual motion machine," whose design is supposed to produce endless work upon being given only the smallest initial effort on one's part. As we'll see later, this is also doomed to failure given the system of rigorous "checks and balances" on energy and work carefully enforced by the universe.

---

[1] On the other hand, simply falling from this same height requires no work at all on your part. In fact, it's the Earth that is doing the work now. This is why going down the stairs is so much easier than going up.

CHAPTER 2

Swinging, Falling, and Rolling

The Initial Foundations of Energy

Our discussion of simple machines shows that nature is not willing to give energy away for free. That is to say, while it's clear these devices make one's life easier (and even today, we continue to use them as the building blocks of much more complex machines that are both human- and fuel-powered), there's a compensatory effect. And as far as we know, there's just no way around it.

As people studied other systems, some simple and others more complicated, this underlying theme continued to reappear in a variety of forms. The next round of insight came from experiments with swinging pendulums, falling objects and objects rolling down – well, what else – an inclined plane (yes, it makes another appearance but not merely as a simple machine this time). These experiments revealed very important results that formed the initial foundations of our understanding of energy. And no one spent more time looking at these systems than Galileo Galilei.

A Swinging Chandelier

Galileo Galilei was born the oldest of seven children in Pisa on February 15, 1564, to Vincenzo Galilei and Giulia Ammannati. Vincenzo, both a music practitioner and music theorist, made only a meager living as a performer and teacher although a number of his compositions were published. From his most significant book, *Fronimo* (containing many compositions for two lutes), we see the true extent of Vincenzo's passion (or addiction) for music: he played his lute "walking in town, riding a horse, standing at the window, lying in bed."

Galileo learned several things from his father. He became an accomplished lute player in his own right as a result of many duets with his compulsive father, in which Galileo played second lute. Like his father, Galileo was a freethinker, and they both enjoyed remarking how people who invoke authority to win arguments are fools. As a firm believer in empirical validation, Vincenzo carried out experiments to support his musical theories. In particular, he established the fundamental relationship between pitch and the corresponding tension of a string: the pitch (or *frequency*) of a vibrating string varies as the square root of its tension. His father's high regard for the necessity of experimental observation to verify theory undoubtedly influenced Galileo, as it would become the cornerstone of all of his scientific inquiries. His mother, although an educated woman, was stubborn and difficult with evidently little affection for Galileo or his younger brother Michelangelo, who remarked a year before she died (in 1620) that he was surprised to learn she was "still so terrible."

Galileo lived in Pisa until he was about ten and then moved to Florence. After some initial schooling from a tutor (who charged five lire a month), he eventually began a more regular curriculum at the mother monastery of the Vallombrosan order near Florence. His time there was most likely responsible for his lifetime romance with astrology. Moreover, it convinced him that he had a religious calling, which his father promptly ended by removing him from the monastery on the pretext that Galileo's eyes were in need of medical attention.

Deciding that Galileo should pursue a career in medicine (probably because it was a prestigious profession that paid well, and because Galileo's most distinguished ancestor was a doctor), Vincenzo enrolled him at the University of Pisa in 1581. In those days, the task of becoming a doctor meant learning Aristotle's natural philosophy by heart. This must have been a frustrating task for Galileo, who wrote, "It seems that there is not a single phenomenon worth attention that he [Aristotle] would have encountered without considering."

Out of all the topics that Aristotle spoke of, it was physics that captured Galileo's imagination. Nonetheless, Galileo approached Aristotle's teachings on the subject with much inquiry.[2] Evidently, Galileo was questioning more than just Aristotle's teachings, as his early years at the university earned him a reputation for contradicting his professors, who he viewed mostly as arrogant, uncritical thinkers who desperately clung to senseless tradition. Galileo began to lose interest in his medical courses. It was during this time that Galileo's life would take a dramatic detour.

Every year around Christmas Grand Duke Francesco's court would move from Florence to Pisa, where it would reside until Easter. Among the entourage was Court Mathematician Ostilio Ricci (1540–1603). In 1583, during Galileo's second year at the university, Ricci was in Pisa, teaching Euclid's[3] *Elements* to the court pages, and Galileo was in attendance. Since these lectures were only open to members of the Tuscan court, Galileo had to remain hidden behind a door as he listened. This was Galileo's first taste of real mathematics, and he was hooked.[4]

He returned to hear more lectures, all the while keeping his presence a secret. Inspired by these lectures, Galileo studied Euclid on his own. Eventually he approached Ricci with questions, and it was then that the

---

[2] Galileo's skepticism of Aristotle's teachings in physics was warranted. Whereas Aristotle contributed substantially to the fields of logic, psychology, political science, and various problems in biology (especially in the classification of plants and animals), little of his work on the physical sciences proved to be of enduring value, save for coining the word *physics*, derived from *phusika*, the Greek word for "nature."

[3] Euclid was a mathematician who flourished around 300 BC.

[4] Most likely Galileo was using the Italian translation of Euclid's *Elements* by Niccolò Tartaglia (1499/1500–1557). The distinction of this text is that it correctly and fully describes the proportion theory of Eudoxus of Cnidus (c. 390 BC–c. 340 BC; Greek mathematician, astronomer, and philosopher), unlike the other two Latin texts existing in Galileo's time. It was Eudoxian proportion theory that enabled Galileo to develop his new science of motion.

court mathematician saw Galileo's gift for mathematics. Ricci encouraged Galileo to continue his self-studies and offered his own help. After an initial introduction by Galileo, Ricci and Vincenzo became friends. Ricci made it known to Vincenzo that Galileo had a talent for mathematics and preferred it to his studies in medicine. Vincenzo (a good mathematician in his own right) had nothing against mathematics, but wanted his son to finish his medical degree. Thus, he agreed to allow Ricci to teach Galileo but thought it best if it was under the guise of being against his wishes in the hopes that Galileo would continue with his usual course work. It didn't work. Galileo completely neglected his medical education and left the university in 1585 without a degree.

After leaving the University of Pisa, Galileo continued to study mathematics on his own, as well as offer private instruction in it in Florence and Siena. During this time, Ricci introduced Galileo to Archimedes' work (c. 287 BC–c. 212 BC). Whereas Euclid's work provided Galileo with a solid mathematical foundation, Archimedes' showed him the power of mathematics when applied to physical problems. Indeed, Galileo was Archimedes' biggest fan and would remain so for life. However, the physics of Archimedes only applied to objects that were standing still. It would be Galileo who extended physics to objects in motion.

In 1586, Galileo wrote his first scientific essay called *The Little Balance*, wherein he described how to build and use a device for measuring specific gravities. This work contained a combination of both pragmatic and theoretical aspects; the latter was taken from Archimedes. In 1587, Galileo discovered a way to determine the center of gravity for certain solids. With this both innovative and practical approach, he had now advanced beyond Archimedes and gained the notice of prominent mathematicians both in Italy and, for the first time, abroad.

Galileo applied for a vacant chair of mathematics at the University of Bologna in 1588. At this time his mathematical grooming consisted of individual sessions with Ricci, teaching private lessons in Florence and Siena, and self-studies. Undoubtedly well aware of his lack of professional experience, Galileo put "around 26" for his age on the application; he was twenty-three. The position was awarded to Giovanni Antonio Magini

(1555–1617). Magini was an astronomer, astrologer, had published some books, and was nine years older than Galileo. It also probably didn't hurt that he was a graduate of the university.

Galileo was becoming known, which, along with help from his patrons, landed him an appointment to the position of professor of mathematics at the University of Pisa in 1589. Galileo now made just half of his predecessor's final salary, making him among the lowest paid of his colleagues at the university. While at Pisa, Galileo managed to offend the professors of philosophy with his criticism of Aristotle's physics, and it was becoming clear that his contract at Pisa would probably not be renewed when it expired in 1592.

Already in 1590, Galileo's friends and patrons began to rally around the possibility of Galileo's getting the chair of mathematics at the University of Padua, which had remained vacant since 1588. It's a testament to Galileo's reputation as a mathematician (in spite of the philosophy professors at Pisa he angered) that in 1592, he was appointed professor of mathematics at the University of Padua, where he now made three times his previous salary at Pisa.

At Padua, Galileo spent his initial eight years getting settled, forming friendships, and setting up shop. He lived a relaxed lifestyle, maintaining his many interests, and focusing his scientific investigations more on the practical rather than the theoretical. In 1599, Galileo acquired a large house with a garden and vineyard. Here he housed students who stayed with him for extended periods (along with their servants) and maintained a workshop (complete with a coppersmith) for the manufacture of instruments. The private lessons he gave, along with his university courses, kept Galileo very busy.

Galileo's most creative years in the science of motion were from 1602 to 1609. During this time, he most likely bounced from one idea to another, allowing both theory and precise experiments to guide the way to his final conclusions that would strike a fatal blow against Aristotelian physics.

In 1583, so the story goes, while attending mass at a cathedral in Pisa, Galileo observed the swinging motion of a chandelier that was put in motion by a passing breeze or perhaps the person who had just lit it. It was clear to him that, if left undisturbed, consecutive swings became smaller and smaller,[5] but how much time did it take for each of these consecutive swings to occur? Using his pulse to measure the time (precise clocks weren't around yet), he was surprised to find that while the size of each swing became smaller, the time for it to occur stayed the same. He was intrigued.

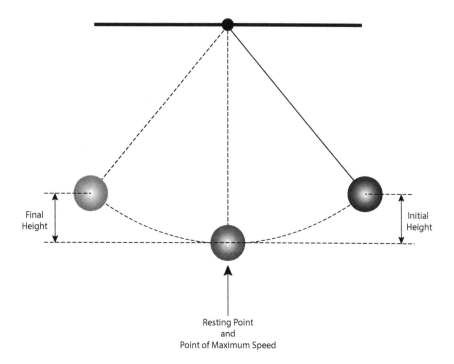

Figure 2.1. A pendulum is moved to the right, away from its resting (and lowest) point (where it's hanging vertically), to an initial height (*amplitude*). Once it's released from here, it swings to the left, through its resting point (where its speed is now at a maximum), to the other side where it reaches its final height (which is the same as its initial height). When the initial height is small, the time it takes for it to swing forward and back again depends only on the length of the string.

---

[5]   Here, the size of each swing becomes smaller because of friction from air and internal friction, such as that between the attached rope and the pivot point. An ideal pendulum would not lose energy to friction, and each consecutive swing would start and end at the same height as the last.

Although it's uncertain if this story is true,[6] Galileo's first notes on the subject of a swinging pendulum (Figure 2.1) – a good model for a swinging chandelier – occurred in late 1588 or early 1589, although he didn't begin conducting experiments[7] until 1602. From his experiments Galileo concluded that the time required (*period*) for a swing of a pendulum does not depend on the size of the swing (*amplitude*); nor does it depend on the amount of mass[8] at the end. Rather, the only thing it depends on is the length of the rope. This means that if you start a pendulum at a height, or an amplitude, that is "small,"[9] then regardless of this initial height, the time

---

[6]   A more likely story is that Galileo noted this type of motion while helping his father conduct experiments of his own involving the tension of musical strings in 1588–1589, and later recalled having previously seen the same motion in the swinging of a cathedral chandelier without ever having considered the physics of it. Indeed, this is what he has Sagredo say in *Discourses on Two New Sciences*:

> A thousand times I have given attention to oscillations, in particular those of lamps in some churches hanging from very long cords, inadvertently set in motion by someone, but the most that I ever got from such observations was the improbability of the opinion of many, who would have it that motions of this kind are maintained and continued by the medium, that is, the air. It would seem to me that the air must have exquisite judgment and little else to do, consuming hours and hours in pushing back and forth a hanging weight with such regularity.

The experiments in question would have involved suspending varying weights from a string, plucking it, and noting the pitch of the resulting sound. Clearly, small swinging motions of the attached weight would occur as an artifact of such experiments.

[7]   This experiment is simple enough to try at home. If you're unable to reproduce Galileo's results, the most likely sources of error are that you're displacing the pendulum too much, you're giving it a slight push before you let it go, the rope is not tight throughout the swing, or the rope is too thick, thus causing internal friction.

[8]   Don't be confused by mass and weight. Weight is simply a measure of the force of gravity acting on the mass of an object. So while weight can change, mass does not change for an object under ordinary circumstances. For example, your weight on the Moon would be less than on the Earth since the gravitational force of the Earth on your mass is greater than that of the Moon. However, your mass would be the same both on the Moon and the Earth.

[9]   This is purely from the mathematics of solving the *equation of motion* for the pendulum. Specifically, it has to do with an approximation one can make when

it takes for the pendulum to swing forward and back to the same position (*oscillate*) will always be the same (within error of the resistance imposed by air and internal friction).

In *Discourses on Two New Sciences*[10] Galileo discusses his thoughts (through his protagonist character Salviati):

> Accordingly, I took two balls, one of lead and one of cork, the former more than a hundred times heavier than the latter, and suspended them by means of two equal fine threads, each four or five cubits long. Pulling each ball aside … I let them go at the same instant, and they, falling along the circumferences of circles having these equal strings for radii, passed … and returned along the same path. This free oscillation repeated a hundred times showed clearly that the heavy ball maintains so nearly the period of the light ball that neither in a hundred swings nor even in a thousand will the former anticipate the latter by as much as a single moment, so perfectly do they keep step.

This observation is only partially accurate, however. Perhaps Galileo's experiments with pendulums involved only small swings, or the clocks he made use of simply weren't accurate enough. While it's true that a pendulum's period of oscillation depends on the string length, rather than the attached mass, if the size of the swing becomes large enough, the period will also depend on the amplitude, or initial height. At this point, the period will become longer as the amplitude increases. Thus, we make the distinction

---

the amplitude – which is measured as an angle from the vertical – is small. The approximation is that the sine of the angle goes as the angle itself (when measured in radians). Mathematically we write: $\sin \theta \sim \theta$, where θ is the angle. This approximation greatly simplifies solving the problem.

[10] Galileo focused on the science of motion until mid-1609. At that time, Galileo became aware of a spyglass (the precursor to the telescope) built by a Dutch spectacle maker in 1608, and he built his own improved version. His focus on his original work on motion occurred once again in 1633 as he began working on *Discourses on Two New Sciences*. In it he recounts his studies on the strength of materials and the motion of objects. Galileo regarded *Discourses* as the best of all his works, a testament of almost thirty years of study by Galileo.

and call a pendulum that swings with a constant period an *isochronous pendulum.*

Galileo considered all pendulums to be isochronous, and this invoked the idea of constructing a reliable clock, something he desperately desired for his experiments. He boasted of having such a clock to the Dutch States General: "These clocks are really admirable for observers of motion and celestial phenomena, and their construction is very simple."

Galileo was bluffing; he didn't have a working model of this "simple-to-construct" clock. However, he did have a theory as to how one might be built, which he elaborated upon with his son, Vincenzo (1606–1649), and his student and first biographer, Vincenzo Viviani (1622–1703). Unfortunately, at the time of his death, Galileo still didn't have a prototype in hand. Eventually, his son did build one in 1649, and in northern Europe it became known from a sketch by Viviani.

In 1656, Christiaan Huygens (1629–1695) had independently come up with a better design. He realized that a pendulum would only keep a constant period for small swings. He was able to overcome this shortcoming by adjusting the pendulum's swing to move along a modified curve, rather than the natural circular one, which maintained a constant period for all heights. This curve is known as a *cycloid*, or *tautochrone.* Huygens published his version of his (cycloid) pendulum clock in 1658 in *Horologium* (meaning *clock* in Latin), and in 1673 published a geometrical proof of the tautochrone as the true constant period curve in *Horologium Oscillatorium.*

The pendulum provides us with other valuable insights. We find that the pendulum's speed at a given point depends on its current height relative to the initial height, resulting in a faster speed when it's at a lower point in the swing. In other words, its current speed depends on the *height difference*: the bigger this difference (the farther away from the starting point), the faster its current speed. Therefore, its maximum speed occurs at the lowest point in the swing, which is also the point at which it will eventually come to rest.

This intimate relationship between height and speed provides us with a further glimpse into the conservation of energy. Galileo would revisit the pendulum again, and when he did he would begin to unravel the mystery of energy even further. However, before we get to this, let's talk about *free fall*.

Free Fall

We learned several things from our discussion of the pendulum:

- The period of oscillation never depends on the amount of mass attached to the end of the rope.
- The pendulum's speed increases as the height decreases, with the maximum speed occurring at the lowest point in the swing.

These results alone are interesting, but they will become more so once we relate them to other types of motion.

A pendulum swinging back and forth is really just an object that has been prevented from making a "complete fall" to the ground due to the attached string. In other words, the string is preventing the pendulum from *free fall*. Think of it like a bungee jumper leaping off the side of a bridge. The first time through he jumps off in the usual fashion with the cord tied around him to ensure that he safely avoids crashing into the ground at the end. Of course, the cord needs to be shorter in length (when fully stretched out) than the jumper's initial height above the ground in order to stop him safely. On his second attempt he jumps again, but now the cord happens to be much longer in length (when fully stretched out) than his initial height. However, rest assured, there's a huge crash pad below to break his free fall, thus ensuring he is not injured.

These are very similar scenarios. The only real difference is the length of the cord: it's shorter than the jumper's initial height on the first jump and longer than the initial height on the second jump. Essentially, this is the relationship between the swinging pendulum and a free-falling object. Therefore, we might expect the physical laws governing both these motions to be similar.

Free-falling objects (Figure 2.2) captured Galileo's interest. Aristotle had said that a heavier object would fall to the ground faster than a lighter one, but Galileo suspected otherwise. Galileo's initial doubts arose when he was a student at the University of Pisa. In a note written several years later, Galileo recalled that his objection was based on his consideration of hailstones of different sizes falling to the ground. Galileo had observed both large and small hailstones hitting the ground simultaneously, rather than the larger arriving before the smaller, as Aristotle would have it. Assuming that both began their fall together (somewhere high above in the sky), Galileo concluded that Aristotle must be wrong.

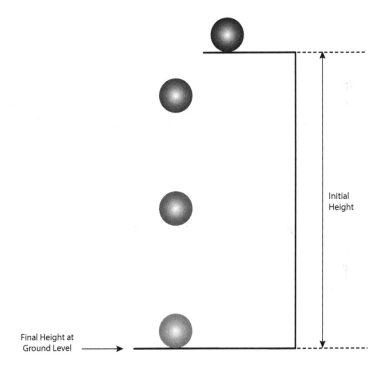

Figure 2.2. An object is pushed off the side of a building (or tower) from an initial height. As it falls, its speed increases (as the height decreases). Its maximum speed occurs just prior to hitting the ground. The time it takes to hit the ground depends directly on the initial height.

Galileo wasn't the first to question Aristotle's theory of falling objects,[11] nor was he the first one to put it to the test.[12] We have it on Viviani's account that during his time as professor at the University of Pisa (1589–1592), Galileo demonstrated the incorrectness of Aristotle's claim by dropping objects of the same material but different weights from the Leaning Tower of Pisa:

> … he quite gave himself over to its study; and then, to the great discomfort of all the philosophers, through experiences and sound demonstrations and arguments, a great many conclusions of Aristotle himself on the subject of motion were shown by him to be false which up to that time had been held as most clear and indubitable, as (among others) that speeds of unequal weights of the same material, moving through the same medium, did not at all preserve the ratio of their heavinesses assigned to them by Aristotle, but rather, these all moved with equal speeds, he showing this by repeated experiments made from the height of the Leaning Tower of Pisa in the presence of other professors and all the students.

Galileo concluded that objects with different weights of the same material hit the ground at exactly the same time; Aristotle was proven wrong once and for all. Well, this is the story, at least, as first told by Viviani in 1657, who recorded what Galileo recounted during his final years. Today, most historians don't believe Galileo actually dropped things from the Leaning Tower of Pisa.

Regardless, we can't help but wonder if Galileo anticipated this result from his studies with the pendulum. After all, we noted before that a

---

[11] The first notable rival theory came from the mathematician and astronomer Hipparchus (c. 190 BC–c. 120 BC) about two centuries after Aristotle's. In 1553, Giambattista Benedetti (1530–1590) was the first to propose and offer a proof for the concept that objects made of the same material with different weights would fall with equal speed through a given medium (e.g., air).

[12] In 1586, Simon Stevin (1548–1620) showed that two bodies of different weights fell at equal speed.

pendulum is simply a modified version of free fall. Therefore, since the period of a pendulum – which also determines its *fall time*[13] (the time it takes to fall to the lowest point in the swing) – is independent of its mass,[14] it should be of no surprise that the fall time of a free-falling object (the time it takes to fall to the ground) is as well.

We find other similarities between a swinging pendulum and a free-falling object. Once again, the speed at any point along the fall depends on the height difference, and the maximum speed still occurs at the lowest point – right before it hits the ground. What about the fall time? We already noted that the period determines the fall time for a pendulum. For the isochronous pendulum this means the fall time, like the period, only depends on the length of the string; that is, it doesn't depend on the initial height (amplitude). However, we also noted that this is really a special case for the pendulum, and in general the period – and therefore the fall time – will depend on the initial height, such that a higher height results in a longer fall time.

This is also true for a free-falling object where the greater the initial height, the longer it takes to fall to the ground. So the key relationship between height and speed is showing up again in free fall as it did for the pendulum. And again it all has to do with the conservation of energy. Let's look at another system, the inclined plane.

Rolling Down Inclined Planes

We already talked about the inclined plane when we discussed simple machines, but now we want to understand the motion of an object rolling

---

[13]   The period of a pendulum is the time it takes for the pendulum to swing forward and then back again to its original position (e.g., left, right, and left again). Here, when we talk of a pendulum's fall time, we simply mean the time it takes for it to fall to the lowest point of the swing. This allows us to make comparisons with the fall times of objects that are free falling and those that are rolling down an inclined plane.

[14]   We focused on the case where the amplitude was small but the principle applies to all amplitudes.

down an inclined plane (Figure 2.3).[15] By now, it should be clear to you that, like the pendulum, this is just another modified version of free fall. Whereas the pendulum was prevented from free falling because of the string attached, the motion of an object on an inclined plane is restricted to only rolling down the slope.

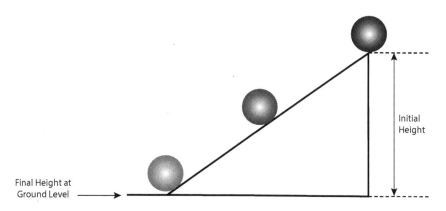

Initial Height

Final Height at Ground Level

Figure 2.3. An object is pushed down an inclined plane from an initial height. As it rolls down, its speed increases (as the height decreases). Its maximum speed occurs at the bottom of the inclined plane. The time it takes to reach the bottom depends directly on the initial height (and the angle). (See footnote 17 for more details.)

Galileo probably began some initial investigations with objects rolling down an inclined plane in 1602, but then, discouraged by the possibility of getting useful results, changed his focus to the pendulum. However, in 1604, Galileo devised a way to measure the accelerating speed of an object rolling down an inclined plane. The experiments that followed provided Galileo with accurate results that he then used to connect to the motions of free fall and a swinging pendulum.

---

[15]   As an object rolls down an inclined plane, its total motion can be separated into the *rotational motion* about the center of mass and the *translational motion* of the center of mass. Even though we are talking about an object rolling down an inclined plane, I am not actually treating the rotational portion of the motion explicitly. Rather, I am focusing only on the object's height change from initial to final, which is an aspect of its translational motion.

Galileo wasn't satisfied to simply know that two objects of different masses fall to the ground at the same time. Rather, he wanted to know how long it took for a falling object to reach a certain height above the ground. To be sure, he was looking for a mathematical expression relating height and time. Unfortunately, Galileo had some experimental challenges to overcome to solve this problem.

While very accurate means of measuring distance and weight existed, this was not so for measuring time; Galileo needed to create a "stopwatch." Galileo's stopwatch consisted of a container filled with water that had a hole in the bottom of it. As the water ran out of the bottom of the container at a constant rate (of approximately three fluid ounces per second), Galileo had an accurate way to measure time. Galileo describes his design and assures of its accuracy in *Discourses on Two New Sciences* (once again, through the character Salviati):

> For the measurement of time, we employed a large vessel of water placed in an elevated position; to the bottom of this vessel was soldered a pipe of small diameter giving a thin jet of water, which we collected in a small glass during the time of each descent ... the water thus collected was weighed, after each descent, on a very accurate balance; the differences and ratios of these weights gave us the differences and ratios of the times, and this with such accuracy that although the operation was repeated many, many times, there was no appreciable discrepancy in the results.

However, even with his water clock, the speed of an object in free fall was simply too fast for Galileo to accurately measure. Instead, Galileo devised a way to slow down the free fall while preserving the key physical results he was after and to allow him to make accurate measurements with the water clock.[16] Galileo's plan was simple and elegant: have the object roll down

---

[16] The late Stillman Drake, Canadian historian of science and expert on Galileo, suggested that Galileo initially used another means of timing objects rolling down the inclined plane. Both Galileo's father and brother were musicians, and Galileo himself was a competent lute player; perhaps Galileo made use of his musical inclinations. Placing rubber bands – or, in Galileo's case, gut "frets" – around the

an inclined plane. The object now "falling" at this slower speed allowed Galileo to perform accurate measurements with his water clock. Galileo was convinced that whether an object rolls down to a certain height (via an inclined plane) or free falls to that exact same height, the underlying physics is still the same. Therefore, he anticipated that the mathematical expression to calculate the time to reach the height – although not exactly the same[17] – was still very similar for both routes. After all, the only difference between an object free falling to a given height and one that is rolling down to that same height is that the latter moves in both the vertical (height) and horizontal (length) directions,[18] whereas the former only has a vertical direction since the object simply falls straight down to the ground.

Inherent in Galileo's assumption was that the vertical and horizontal directions of the object's motion down the inclined plane are independent

---

inclined plane would have caused a sound to be made each time an object rolled over one. Adjusting the placements of these frets such that the object reached each of them at the same interval of time would then allow an accurate determination of the time. In order to set such a placement, Galileo could have started the object rolling, listened for the "pluck" of each fret, and then using his internal sense of rhythm (perhaps he tapped his foot, or sang a marching tune), made the necessary adjustments in position until each fret represented a set interval of time. Now all that was needed was to measure the distance from the initial position of the ball to each of the frets, which was something Galileo could determine very accurately. Given that the clocks of the day couldn't measure times shorter than a second, this method would have most likely been more accurate.

[17] In both cases, the general form of the mathematical equation relating time and height is the same, and is given by $h = \frac{1}{2}at^2$, where h is the height the object has fallen (or rolled) to from its starting point, a is the acceleration, and t is the time. In other words, the height goes as the time squared. This is *Galileo's Law of Fall*, and he figured it out from data obtained from his experiments with the inclined plane. For free fall, the acceleration is due to gravity, which is 32 ft/s^2, giving $h = 16$ ft/s$^2 \times t^2$. However, for an inclined plane the acceleration is slower than for free fall. Moreover, for the inclined plane the acceleration depends on the angle of inclination, which is the angle the slope of the inclined plane is making with the (horizontal) ground; for free fall the angle of inclination is 90°.

[18] Therefore, we need both these directions to fully describe the motion of the rolling (falling) object.

of each other and, therefore, they can be treated separately. This would mean that regardless of the horizontal direction, the laws of physics governing the vertical direction (the one he was most interested in) are the same for free fall and rolling down an inclined plane. Well, it turns out that Galileo's hypotheses were right.

By now, it should be no surprise to you that the speed of an object rolling down an inclined plane[19] increases with the height difference, the maximum speed occurs at the bottom, and the fall time (the time it takes to roll to the bottom of the inclined plane) doesn't depend on the mass but is directly related to the initial height, just as it was for (the general case of) the pendulum and for a free-falling object.

So, for all three systems the results are the same because of nature's requirement that energy always be conserved. Now, it's true that we haven't discussed in detail what this conservation of energy really entails; I've been holding out a bit. Nonetheless, for the systems discussed, we have two fundamental relationships between height and speed:

- A lower height (from the starting point) means the object is moving faster, which means its fastest speed is at the lowest point.
- The higher the initial height, the longer the fall time, except for the isochronous pendulum, which has the same fall time for every height.

Let's take a look at another variation of Galileo's pendulum experiment.

---

[19] It's interesting to note that in *On Motion*, which documents Galileo's scientific work during his professorship at Pisa (1589–1592), Galileo thought that the speed of an object on an inclined plane went inversely as the length of the slope. Indeed, his inability to correctly account for the motion on an inclined plane (since he didn't recognize the importance of the acceleration due to gravity at the time) is most likely why Galileo never completed and published *On Motion*. However, by the time he wrote *Discourses on Two New Sciences* Galileo had figured out the correct relationship: for constant acceleration, the speed, v, of a falling object is directly proportional to time, t. In other words, $v \sim t$. Therefore, the speed of a falling object will increase as the height decreases.

The Pendulum Revisited

In his experiment with the "interrupted pendulum," Galileo uncovered more implications of the conservation of energy. Recall that Galileo's pendulum was simply a lead ball weighing one to two ounces and attached to a fine piece of thread. Now, imagine a pendulum suspended from a nail driven into a wall, and it can swing freely from one side to the other (Figure 2.4). From its resting position (simply hanging vertically), we move the pendulum to, say, the right to some initial height and then release it, without giving it a push.[20]

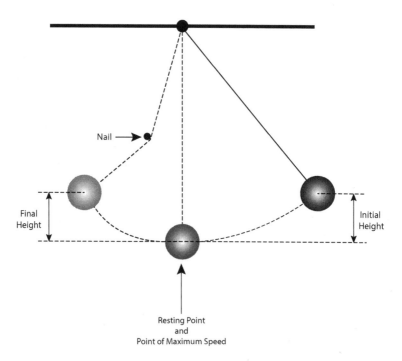

Nail →

Final Height

Initial Height

Resting Point
and
Point of Maximum Speed

Figure 2.4. As before, a pendulum is moved to the right, away from its resting (and lowest) point (where it's hanging vertically), to an initial height and released. As it swings to the left, the string catches on a nail, which forces it to alter its path. Regardless, the pendulum still reaches a final height that is the same as its initial height.

---

[20] Pushing it will change the described results since you're now adding more energy into the system.

As the pendulum swings from right to left, we find that it arrives at its final height. Galileo probably did this many times using different initial heights, and each time he found the same result: the initial height always equals the final height. Well, to be honest, the final height was probably a bit lower due to some air resistance, but Galileo deduced that neglecting this would result in equal heights, which is the key result.

But then Galileo added a twist to the original experiment. Now imagine the exact same setup as before, except that this time we drive a nail into the wall such that the string will catch on it as the pendulum swings from right to left. Although the swing of the pendulum is altered due to the nail, we find, once again, that the initial height and the final height are equal. What if we change the position of the nail? It doesn't matter. The string will simply catch on the nail, the swing will be altered, and the pendulum will arrive at its final height, which (as before) will be the same as the initial height.

Let's consider one last possibility: What if the nail makes it impossible for the pendulum to alter its swing such that it can actually reach a final height that is equal to the initial height? In that case, the pendulum simply continues its motion as it wraps itself around the nail.

When we talked about the pendulum before, we learned that as it swings down, getting further from its starting height, its speed actually increases. In other words, the decrease in height results in an increase in speed. Now, we find that as the pendulum continues through on the upswing, its final height (or maximum height) will be the same as the initial height. How are these concepts related? It turns out that the interplay between the height and speed are exactly balanced. We learned before that the gravitational force on an object at a given height imparts potential energy to it, but we never talked about its counterpart, which is *kinetic energy*. Whereas the potential energy is "stored-up energy," kinetic energy is "energy of motion" that gives an object its speed.

We discussed earlier how work is conserved such that a decrease in required force results in an increase in the distance over which it must be applied when using a simple machine. Nonetheless, the overall work to do the task is conserved.

Kinetic energy and potential energy share a similar conservation relationship. In the case of the pendulum, it means that as the height decreases, the loss of potential energy is exactly compensated by an increase in kinetic energy, which means the speed increases. And vice versa: as the pendulum continues through its upswing it gets closer and closer to its starting height (via the other side), and accordingly the kinetic energy decreases, the pendulum slows down and rests for a moment at its final height (which is the same height that it started at) before falling back down. Therefore, the pendulum is moving at its fastest at the lowest point in the swing, while its slowest speed occurs at the very top of the swing. This exchange between potential energy and kinetic energy isn't unique to the pendulum; it applies to all systems (inclined plane, objects in free fall, and others) and is in perfect balance when *friction*[21] is absent.

By mid-1609, Galileo was working on his treatise about the science of motion, and upon hearing of the invention of the spyglass (the precursor to the telescope) he dropped everything to make his own version. By the end of August, Galileo had a 9X telescope, which he presented to the Venetian Senate and dignitaries. For his efforts, they rewarded him by doubling his salary and giving him tenure for life. However, there were some misunderstandings that Galileo learned about after his acceptance. The salary increase wouldn't occur until his existing contract expired, no further increase in salary would ever occur, and he was to remain teaching at the University of Padua for life. Unhappy with this arrangement, Galileo managed to strike a new deal in 1610 whereby he became Chief Mathematician of the University of Pisa and Philosopher[22] and Mathematician to the Grand Duke of Tuscany. The appointment was for life and he wasn't obligated to teach at the university. He also wasn't required to reside in Pisa, which allowed him to finally return to his beloved Florence.

Around December 1, 1609, Galileo had in his possession a 20X telescope, allowing him to observe the Moon's rough, mountainous surface,

---

[21] Friction is the resistance that occurs between two objects in contact with each other that opposes their relative motion. For example, when you drive your car really fast around a corner and don't slide off the road, you can thank the friction between the road and your tires for that. However, if you do slide off, it's because the friction wasn't enough to cause your car to resist this motion and hold you on the road; you should slow down next time.

[22] To be sure, by "philosopher" Galileo meant "physicist."

four (of the sixty-seven currently known) moons of Jupiter, and several new stars. These, along with Galileo's other celestial discoveries, would come to prove the theory put forth by Nicolaus Copernicus (1473–1543)[23] that the planets revolve around the Sun, and Galileo told everyone all about it. Unfortunately, Copernican theory was in direct conflict with the teachings of the Church, which insisted that the Earth was the center of the universe and all the planets revolved around it.

On February 26, 1616, the Church told Galileo that he was:

> ... to abandon completely ... the opinion that the sun stands still at the center of the world and the earth moves, and henceforth not to hold, teach, or defend it in any way whatever, either orally or in writing; otherwise the Holy Office would start proceedings against him.

Galileo agreed to behave. Nonetheless, in 1624, the Pope assured Galileo that he could write about Copernican theory just as long as he approached it strictly as a mathematical proposition. However, in 1633, upon publishing *Dialogue Concerning the Two Chief World Systems*, Galileo, now nearly seventy years old and in very poor health, found himself facing the Inquisition once again. Pope Urban VIII found Galileo guilty of heresy and sentenced him to guarded house arrest. In spite of this and the death of his beloved daughter in 1634, Galileo returned to his project of some twenty-five years, taking three years to produce his final masterpiece *Discourses on Two New Sciences*, completed in 1637 and published in 1638 only after the manuscript was smuggled out of Italy and into Holland.

Surprisingly, the Church didn't punish Galileo further (after all, his book completely overturned Aristotle's physics, which was the only physics the Church supported). Perhaps the spiritual leaders of the Church had been swayed by Galileo's preface of the book indicating that it had been published by his friends abroad without his consent or knowledge, and that he had sent it to them merely for its scientific interest.

---

[23] Copernicus' book was published just before his death.

CHAPTER 3

Untangling the Mess

Energy, Momentum, Force, and Matter

From Galileo's description of his experimental results with pendulums and inclined planes, and even in his thought experiments, it's clear that he possessed a very keen intuition about energy and its conservation. While Galileo detailed his findings in *Discourses on Two New Sciences*, he never fully realized that he had the beginnings of a conservation law for energy. Indeed, he had witnessed firsthand the conservation between kinetic energy and potential energy, which together constitute *mechanical energy*; from his experiments these were the only two forms of energy he knew existed.

Strictly speaking, in the systems Galileo was considering, the mechanical energy would only be conserved in the absence of friction. In his experiments, Galileo strived to eliminate friction and deliberately ignored it when drawing his major conclusions. And by "ignore" we don't mean in the careless sense. On the contrary, Galileo was very concerned with the precision of his experimental measurements.

However, he didn't let his concern for the details prevent him from concluding that certain discrepancies or apparent contradictions were simply things that warranted neglect in order to see the bigger picture. So while Galileo's contemporaries agonized over such details, unable to take the next big step, Galileo firmly believed in the mathematical consistency of nature and left them behind to fret. This ability of Galileo to use his observations from studying real systems (like an object rolling down the inclined plane), where friction was present, and extract a fundamental physical principle illustrates his true genius.

Today, we understand the results of all the systems Galileo studied in terms of the conservation of mechanical energy. Consider an object sitting on your coffee table – say, the remote control for the TV. Now, naturally it has no kinetic energy since it's not moving (I hope). But let's consider this scenario: suppose you ever so gently nudge it to the edge of the table until it finally slips off and (free) falls to the ground. Clearly, as it was falling, it had kinetic energy (until it hit the ground and stopped). But before you bumped it to the edge and caused it to fall to the ground, when it was just sitting on the table totally undisturbed, it had *potential energy*.

In this example, the potential energy has to do with the fact that the remote had the potential to be put into motion once it slipped off the table. Once in motion, the potential energy it had while sitting on the table was transformed into kinetic energy; this is the relationship between the potential and the kinetic energy of an object.

Regardless of what the object is – a swinging pendulum, something rolling down an inclined plane or falling from a building, or simply a remote on your coffee table – the height of the object from the ground gives it potential energy, while falling from this height converts the potential energy into kinetic energy. The potential energy in these examples is coming from the gravitational force of the Earth, which is "pulling" downward on the given object. Without some sort of force "pulling" or "pushing," an object has no potential energy.[24]

While Galileo's experiments dramatically progressed our understanding of mechanical energy, he never actually possessed a clear definition of what *energy* really was. In this respect, he was not alone. The confusion about the definition of energy – both physically and mathematically – was

---

[24]   Another example is a spring, which has potential energy when it's compressed since in this state it's trying to "push itself" apart.

still intimately tied to those for *momentum* and *force*. And if this wasn't confounding enough, it was evident that the nature of matter was somehow intertwined as well; unfortunately, a clear physical understanding of this was even further away. Galileo died in 1642, his last nine years spent under house arrest and the final four years of his life leaving him in total blindness. Suffice it to say that Galileo came further than anyone had up until then in describing energy. At the time of his death, the true nature of energy was still a mystery, and a complete understanding would have to wait for more than two hundred years.

When Things Collide

By the late seventeenth century, mathematics was beginning to provide powerful tools for the description of physical phenomena. Despite having the needed mathematics, the confusion around energy, momentum, and force persisted for a long time.

In part, this was because the key players didn't all have the same mathematical training, ability, or style. At this time, mathematics wasn't just for the trained professional. Aristocrats and educated people also found it fashionable to dabble, and outsiders hoped to stake their claim to fame by contributing to academic competitions (complete with prizes). Mathematics wasn't just a tool for solving physical problems; it also provided a means for furthering one's career, building alliances, establishing influence, and impressing others.

Aside from the politics, there were other (more important) points of contention centering on the physical interpretation of the resulting mathematical quantities. Specifically, what do the *physical properties* of energy, momentum, and force look like from a *mathematical perspective*? Moreover, of these physical properties which, if any, were conserved?

The idea of certain properties being conserved began to take on importance as scientists and mathematicians became increasingly convinced (often based on intuition, metaphysical, philosophical, or religious reasons, rather than scientific reasoning alone) that conservation was something fundamental in the way the universe worked.

As a result, the idea of conservation started being used more often in mathematical calculations[25] (sometimes incorrectly), which also gave it more visibility. The main topic of these intense discussions often focused on the head-on collision (*impact dynamics*) of "hard spheres," such as the collision between two billiard balls on a pool table.

This, in turn, presented yet another issue to address: To what degree can an object be "compressed" or "squished"? In other words, how "hard" are the colliding objects, and can they be deformed from the impact of the collision? Clearly, this speaks to the very nature of matter itself.

Conservation of "Motion"

In 1644, in his *Principles of Philosophy*, René Descartes (1596–1650) proposed that the total *motion* of the universe is conserved. In this way, when two objects collide with each other their combined motion before and after the collision is the same. His basis for this was simple: God made it this way.

> It is obvious that when God first created the world, He not only moved its parts in various ways, but also simultaneously caused some of the parts to push others and to transfer their motion to these others. So in now maintaining the world by the same action and with the same laws with which He created it, He conserves motion; not always contained in the same parts of matter, but transferred from some parts to others depending on the ways in which they come in contact.

According to Descartes, the motion of an object was correctly measured by the quantity $m|v|$; in other words, the mass, m, of the object multiplied by its speed, $|v|$, quantifies the motion. Descartes also introduced a few rules (seven, to be exact) to allow one to correctly predict the outcome of an isolated collision between two "perfectly solid" bodies. His rules were in direct contrast to everyday experience, and Descartes acknowledges this:

---

[25] The conservation of energy is a fundamental and often necessary tool in solving many physical problems.

Indeed, experience often seems to contradict the rules I have just explained. However, because there cannot be any bodies in the world that are thus separated from all others, and because we seldom encounter bodies that are perfectly solid, it is very difficult to perform the calculation to determine to what extent the movement of each body may be changed by collision with others.

Nice sentiment, but wrong.

Momentum, Not "Motion"

In 1666, an experiment with colliding bodies drew much attention from the Royal Society, and Robert Hooke (1635–1703) began demonstrating his own collision experiments at the weekly meetings. Others had begun experiments as well, among them Christopher Wren (1632–1723) who, along with Christiaan Huygens (1629–1695) and John Wallis (1616–1703), was invited in 1668 to present a theory on the related laws of motion. This was twenty-four years after Descartes' *Principles of Philosophy*.

Shortly after submission, the articles were read before the Society: Wallis' on November 26, 1668, Wren's on December 17, 1668, and Huygens', published later that year in both *Philosophical Transactions* and in the *Journal des Sçavans*, on January 7, 1669. Huygens' submission was taken from the work he had already completed in 1656, which he had decided not to publish at that time; it was published posthumously in 1703 as *De Motu Corporum ex Percussione*.

Independently, they had each concluded that it's the total *momentum* during the collision that is conserved, not the total *motion* as Descartes had insisted. Unlike Descartes' quantity of motion, $m|v|$, the momentum of an object is given by mv, that is, the product of the mass multiplied by the velocity, v – not the speed, $|v|$. What's the difference?

Imagine that you're driving down the road in your car and you look at the speedometer, which gives your speed $|v|$; this was the quantity that Descartes was referring to. Now imagine that you, once again, look at the speedometer and then at a compass; now you know both your speed and direction; this is velocity, v. We call a quantity like v a *vector*, since it indicates both direction and magnitude, whereas a quantity like speed, $|v|$, which we can more simply write as v, gives only the magnitude and is called a *scalar*.

Huygens, who had developed the more complete theory, went one step further and concluded that for "hard spheres" that collide with each other and recoil to their original pre-collision form, the quantity mv^2 for the colliding objects is also conserved. We now call these types of collisions *elastic*, as opposed to *inelastic* collisions, where the colliding objects suffer some sort of deformation – they get "squished" – that persists after the collision.

The work of Wallis (who also considered inelastic collisions), Wren, and Huygens went a long way to improve the understanding of impact dynamics. It establishes the initial foothold for momentum conservation and dispelled Descartes' conservation of motion theory. Moreover, Huygens' conservation of mv^2 introduced a new notion of conservation, later helping to revive the argument.

Vis Viva: The "Living Force"

In 1686, Gottfried Wilhelm Leibniz (1646–1716) published his *Brief Demonstration of a Notable Error of Descartes and Others Concerning a Natural Law, According to Which God Is Said Always to Conserve the Same Quantity of Motion; A Law Which They Also Misuse in Mechanics*. In it he argues against the late Descartes' conservation of total motion and provides examples of where the concept fails. Thus began the famous dispute known as the "*vis viva* controversy."

In 1695, in his *Specimen Dynamicum*, Leibniz publicly introduces what he considers the key quantity, mv^2, which he called then *vis viva*, or "living force," the same quantity Huygens had said was conserved in individual collisions between hard spheres. However, for Leibniz, *vis viva* was conserved also universally.

As the living force, Leibniz saw *vis viva* as a measure of an object's ability to impart effect by mere virtue of its motion. In this way, a moving body colliding with a stationary one imparts "life" by putting it into motion. And in general, colliding objects were imagined to transfer *vis viva* between each other without loss, thus enforcing its conservation.

Nonetheless, Leibniz was concerned (and rightfully so) about those collisions that "appear" to lose their *vis viva*. A classic example of this is a falling object colliding with the ground, where it ultimately stops dead in its tracks. Another good example would be two "soft objects" moving towards each other, being "squished" or deformed upon impact and each slowing down in the process.

Leibniz was a smart guy and was well aware of the contradiction. According to him, the *vis viva* isn't lost; it's simply passed to smaller pieces within the object. The catch is that these smaller pieces absorb the *vis viva* but don't contribute to the overall motion of the object, and in this way the *vis viva* is still conserved. He said, "But this loss ... does not detract from the inviolable truth of the law of the conservation .... For that which is absorbed by the minute parts is not absolutely lost for the universe ...."

While Leibniz envisioned *vis viva* as giving an object motion, he viewed *vis mortua*, or "dead force," as giving an object the tendency towards motion. For example, an object sitting on a table has a tendency towards motion by virtue of its height from the floor and its weight, which is realized once it's bumped off the table – it falls to the ground. Therefore, *vis mortua* can turn into *vis viva* as a previously motionless object overcomes whatever is restricting it; once the object is pushed to the edge of the table, it has nothing left to resist its motion of falling to the floor. These ideas were of course the precursors to what we now call potential and kinetic energy and the idea that energy can actually be converted from one form to another: *vis mortua* changing into *vis viva*.

To be sure, Leibniz's conviction of the conservation of *vis viva* is rooted in the metaphysical. Like many natural philosophers, he was simply convinced that nature conserved "something." By having this something be *vis viva*, it assured Leibniz that the universe was a self-sustaining system that would stay in motion, and not some "clock" that would run down if not for constant rewinding.

Leibniz's *vis viva* and the related concepts were a significant step forward. Nonetheless, the very fact that he saw it as an actual measure of force, rather than a form of energy (the modern-day expression for kinetic energy is $\frac{1}{2}\,mv^2$), illustrates the confusion in the contemporary thinking. The *vis viva* argument that Leibniz started in 1686 lasted for some time and wasn't fully resolved until some undetermined date in the eighteenth or possibly even the nineteenth century.

The Laws of Motion

In 1687, Isaac Newton (1643–1727) published *Principia*, which without a doubt is one of the most important works in physics of all time. Although it was remarkably unreadable because of both its content and form, it sold out quickly. In it Newton describes (among other things) his three laws of motion. These three laws established – both mathematically and physically – what a force really is and what it's not. For one thing, it's not energy. Newton goes on to correctly derive the conservation of momentum from his third law.[26] Unlike Huygens, Newton proved that momentum is conserved universally, not just during a collision involving two hard spheres. Today, we find momentum to be conserved in all sorts of systems, from billiard balls to subatomic particles.

Although Newton refrained from engaging in the whole *vis viva* controversy, he was against *vis viva* altogether since it seemed not to be conserved in inelastic collisions; apparently he wasn't convinced by Leibniz's argument of *vis viva* being passed to smaller pieces within the object. In general, Newton simply didn't believe in the conservation of energy. Whereas Descartes was content to have God play the role of the universal architect, Newton – being the very religious guy that he was – required a more prominent and permanent role for God than simply that of an initial designer.

---

[26] It's interesting to note that Newton's derivation of conservation of momentum from his third law results in a strong conservation principle being derived (momentum conservation is never violated) from a weaker one (forces are not always "equal and opposite").

For him, God's involvement was necessary to keep everything running smoothly. Accordingly, energy wasn't conserved because it was God who continued to provide the universe with energy as necessary; it was God's hand that wound up the "universal clock" to keep it running forever. The fact that the universe still continues to run was, for Newton, proof of God's existence. So Newton's universe was one that ran according to his laws of motion and conserved momentum but required an occasional "nudge" from God to keep things on track.

For the most part, Newton had written *Principia* to solve problems involving celestial objects, like the motion of the planets around the Sun, giving only a few examples of how one would apply these laws to terrestrial motion back here on Earth.[27] And while these results were very impressive, the small number of examples and the lack of precision of the derivations left many wondering just how to apply his laws of motion to terrestrial problems in general. In particular, impact dynamics, to many, seemed not to be readily treatable by Newton's methods.

Understanding Matter

The lack of a clear understanding of the composition of matter and its fundamental behavior complicated things even further in humanity's quest to comprehend motion. In 1724, the Paris Academy held a competition concerning the laws governing the collision of "perfectly hard spheres." Up for the challenge, Johann Bernoulli (1667–1748) started by outright rejecting the existence of such bodies in nature. Needless to say, he didn't endear himself much to the academy with this stance, and his entry was disqualified.

For Bernoulli (who based his argument on *Leibniz's Law of Continuity*), if two perfectly hard spheres were to collide, their directions and speeds would have to change instantaneously upon impact. This is because a

---

[27] These included some scattered observations involving the pendulum, the tides, and – with the help of some ad hoc assumptions – a successful derivation of *Boyle's Law*, and a formula for the speed of sound in air.

perfectly hard object won't be "squished" or "compressed," but rather left unchanged from the collision. On the other hand, an *elastic* object will be compressed upon impact, and then subsequently restore itself to its original state afterwards,[28] very much like a spring can be compressed and then expand again. As one could anticipate, this process would need a certain amount of time to actually occur. For Bernoulli and Leibniz, the absence of such a mechanism for perfectly hard spheres is why the collision would have to occur instantaneously and therefore wasn't physically realizable.

Bernoulli saw matter as fundamentally elastic, where a collision resulted in the compression and expansion of "tiny springs" within the impacted objects. As one of these springs is compressed, *vis mortua* is created, and its subsequent expansion leads to its conversion into *vis viva*. In turn, this *vis viva* is then passed to the other body involved in the collision, thereby altering its motion. Using this ingenious model of matter, Bernoulli was able to show, as Huygens had done, that both *vis viva* and momentum are conserved in collisions. In this way, Bernoulli provided a mathematical and physical foundation for Leibniz's original paradigm, thereby extending it.

By the time the eighteenth century came to a close, the area of physics that today we call *classical mechanics* had really come into its own. Galileo had shown that one could understand the universe through careful observation and mathematics. Others continued to build upon the solid foundation that he laid. Mathematics had become even more powerful since Galileo's time, and its application to physical problems more prevalent. Galileo's work had provided an initial basis for the conservation of "something," which eventually turned out to be identified with the conservation of mechanical energy, or in other words, the conversion of potential energy to kinetic energy, and vice versa.

This realization came out of the attempt to better understand momentum, force, and matter, as well as energy. While he refuted the conservation of energy, Newton did prove the universal conservation of momentum, give

---

[28] Sometimes this is called "perfectly elastic."

a both mathematical and physical description of force, and provide the laws of motion for all things celestial and terrestrial.

The work of others would further clarify and extend his efforts. Even a working definition of matter as comprising "tiny springs," although far from complete, had shown success in solving physical problems. Indeed, it seemed like things were pretty much "under control." Nonetheless, there were still many unresolved issues. Of these, perhaps one of the most confounding was heat.

CHAPTER 4

The Missing Link

Heat: The Final Piece in the Energy Puzzle

The majority of our knowledge of energy can be attributed to two areas of physics: *classical mechanics* and *thermodynamics*. Experiments in mechanics involving swinging pendulums, balls rolling down inclined planes, and objects falling off of buildings gave tremendous insight into the properties and behavior of energy, but the picture was still very incomplete. It took studies in thermodynamics to find the last piece of the puzzle, the one that remained so elusive for so long: heat.

When all was said and done, it took the discoveries made over thousands of years in mechanics and those made over several hundreds of years in thermodynamics for us to really understand energy and its fundamental nature. After all the time and effort were expended, a paramount truth emerged: energy is neither created nor destroyed; rather, it's seamlessly transformed from one form to another.

Caloric Theory

Hints of energy being conserved, much like momentum, were showing up by the 1840s. But unlike momentum, which by comparison was quickly accepted and understood, energy still remained a mystery. It was understood that energy could be potential or kinetic, with the former "transforming" into the latter; thus, that a conservation principle in this specific case must undoubtedly exist seemed certain. But this was a far cry from a complete understanding. The biggest remaining piece of the puzzle that eluded explanation was heat.

Systems such as objects rolling down inclined planes and swinging pendulums (your old favorites by now, I'm sure) were well described by the theory of mechanics formally laid out in Newton's *Principia*. These types of physical problems were once solved through the visual relationships of geometry (as Galileo did), but were now described with the somewhat more abstract but much more compact and powerful equations of analytical geometry (as created by Descartes) and calculus (as created by Leibniz and Newton independently of each other).

These new mathematical tools allowed the physical problems of mechanics to be more readily solved with brilliant success. Relationships between actual physical quantities (like force and momentum) and the mathematics describing them were being firmly established along with their experimental counterparts; so one could write down the mathematical equations describing the physical system and then validate this theory in the lab. Indeed, the physics of mechanics was a huge success. But where did heat fit into this remarkable new framework, if it had a place in it at all?

By the end of the eighteenth century, heat – along with its cohorts: light, magnetism, and electricity – was regarded as an *imponderable fluid*. These imponderable fluids were set apart from the (only slightly better understood) "ordinary matter" that constitutes everyday objects as lacking of any definite structure. They were regarded as some sort of fluid, capable of flowing much like water, thus affording them the ability to move freely between the assumed spaces that must be present in ordinary matter, like sunlight passes through the glass of a window, or heat passes through a coffee cup to your hand.

Needless to say, eighteenth-century theories describing physical phenomena were very qualitative whenever imponderable fluids were involved. This was in sharp contrast to the physical problems of mechanics that were described by elegant mathematics.

Pierre-Simon Laplace (1749–1827) imagined heat to be a fluid composed of particles that Antoine Lavoisier (1743–1794) deemed as "caloric." Whereas the particles constituting ordinary matter were considered to be attracted to each other, caloric particles were considered to repel each other.[29]

The fact that the particles of ordinary matter were attracted to each other seemed to be consistent with experimental results: cooling a gas results in the particles moving towards each other to form the more compact liquid structure, and subsequent cooling moves the particles even closer together, resulting in the solid structure when something freezes. By contrast, the addition of heat to a substance meant you were adding particles of caloric to it, and since caloric particles repelled each other, increasing the caloric meant lessening the attraction between the particles of ordinary matter. Therefore, add enough heat to a solid substance (such as ice) and it will melt; add even more and it will boil. Caloric theory seemed to make sense when it came to these phase transitions.

In 1789, Lavoisier published *An Elementary Treatise on Chemistry*, in which he describes thirty-three elements. The list begins with (what else) caloric and continues with light, oxygen, nitrogen, and hydrogen. Lavoisier also discusses his findings from studying a variety of chemical reactions. In particular, he notes that for the chemical reactions he studied, the mass of the starting materials (reactants) equals the mass of the final materials (products).

In other words, regardless of the chemical reaction that is being performed, the total mass of all materials involved will be conserved throughout the course of the reaction. In fact, Lavoisier was able to prove this by simply "weighing" the reactants and products with a very accurate chemical balance that he had constructed himself. He concludes that in general this is a fundamental property of all the elements and therefore, since caloric made his list of elements, it too must be conserved as far as he was concerned.

---

[29] By today's standards, it may seem strange to describe heat as a type of particle, but light – another imponderable fluid – was also being promoted as a particle, especially by Newton himself.

That heat was conserved and therefore couldn't be created or destroyed was central to the concept of caloric theory. In pragmatic terms, it meant if one object lost caloric (heat), another (nearby) object acquired that exact same amount of caloric (heat). In a similar fashion, Lavoisier also found that caloric (heat) is weightless. So although caloric is imagined to be a "material" substance that is conserved, it has no weight. This unsurprisingly caused suspicions in some people's minds.

Although caloric theory seemed to be able to explain some things as already described, there was one big question in many people's minds: How does heat from friction figure into the theory? We are all aware that rubbing two materials together will produce an amount of heat resulting from the friction between them. According to caloric theory, the heat produced is a result of the caloric from one object being "bumped off" by the other object. In other words, when two objects are rubbed together and generate heat due to the friction between them, caloric theory says it's because you're knocking off particles of caloric in the process. Not everyone was buying this explanation.

Heat Is "Motion"

Count Rumford of Bavaria, born Benjamin Thompson (1753–1814), gained insight into the nature of heat while manufacturing cannons in Munich.[30] Rumford noted that when the hole of the cannon was made by boring into it with a drill, the barrel of the cannon became hot. This was no surprise since the process of drilling produces friction between the drill bit and the barrel of the cannon. But what really struck him was the fact that as long as he continued to drill, he could continue to produce heat. That is to say, as long as there was friction (in this case between the drill bit and the cannon barrel), there was always heat being produced.

According to caloric theory, however, heat from friction resulted from the drill bit "setting free" particles of caloric that were stored up in the

---

[30] As major general and commandant of the police to the court of the Duke of Bavaria, Rumford was responsible for the defense of Munich.

cannon barrel all along. If this were indeed true, the caloric should be running out at some point rather than continuing to be produced. Moreover, if there's this much caloric actually stored in the cannon, it should have already melted the canon barrel or at least have made it extremely hot to the touch. This would seem to suggest that caloric was not being conserved (as caloric theory demanded) – evidently it was available in unlimited quantities!

In 1798, Rumford states:

> It is hardly necessary to add, that anything which any *insulated* body
> … can continue to furnish *without limitation*, cannot possibly be a
> *material substance*; and it appears to me to be extremely difficult, if
> not quite impossible, to form any distinct idea of anything capable
> of being excited and communicated in the manner Heat was excited
> and communicated in these experiments, except it be *motion.*

To be sure, in Rumford's mind the motion of the drill boring is "communicated" to the cannon barrel, resulting in some sort of "motion" within it that, in turn, manifests itself as heat. Thus he refutes caloric theory and its notion of heat being a material substance, as Lavoisier proposed. Moreover, he alludes to why heat is indeed weightless as seen in Lavoisier's experiments: it's simply an artifact of some sort of "motion" or "excitation" within, in this case, the cannon barrel.

But if heat is "motion," what exactly is moving inside an object that causes this heat? Unfortunately, Rumford provides no explanation, concluding simply:

> I am very far from pretending to know how, or by what means, or
> mechanical contrivance, that particular kind of motion in bodies,
> which has been supposed to constitute heat, is excited, continued,
> and propagated ….

Thus, without a deeper understanding of the "motion," Rumford's major insight into the nature of heat remained largely ignored until the 1840s.

The Mechanical Equivalent of Heat

The moment Rumford said heat is "motion," he, perhaps inadvertently, connected heat to *mechanics*. In mechanics, motion in terms of objects rolling down an inclined plane or falling from an initial height had been studied extensively by Galileo and was well described by the mathematical equations set forth in Newton's *Principia*. Therefore, if heat were indeed "motion" as Rumford suggested, it seemed plausible that it might fit within the solid framework of mechanics. Among other things, mechanics had been successful at understanding energy in the forms of kinetic and potential; perhaps now it was ready to provide a clearer understanding of heat.

While serving as a doctor on a Dutch merchant vessel to the East Indies, the German physician Julius Robert Mayer (1814–1878) noticed that, upon performing bloodletting, venous blood was brighter red in the tropics than in Europe. Venous blood is blood carried in the veins on its return to the heart. With the exception of the pulmonary veins, this blood is low in oxygen and high in carbon dioxide, having released oxygen to and absorbed carbon dioxide from the tissues. By contrast, the brighter red color of the venous blood observed by Mayer meant that in the warmer climate of the tropics, the body does not use up oxygen as quickly.

Mayer concluded that the human body uses less oxygen in warmer conditions since it needs to burn less fuel to maintain normal body temperature. Further, he concluded that performing physical work and the production of heat to maintain normal body temperature both require food consumption and are in some way equivalent: heat and work are two versions of the same thing.

In 1841, six months after his return home, Mayer sent an account of his ideas to a respected scientific journal, only to have it declined for publication. His thoughts were presented in a very obscure and confused manner, perhaps partly attributable to his lack of knowledge of physics and mathematics. Realizing his limitations, he enlisted a friend to teach him mathematics and physics and submitted an improved version of his manuscript in 1842 to *Annalen der Chemie und Pharmacie*, a journal for chemistry and pharmacy.

The theory was improved in this version, but his metaphysical style persisted. Nonetheless, he provided a value for the *mechanical equivalent of heat.* That is to say, having previously speculated that *work* (as defined in the first section) and heat are different forms of the same quantity, he tried to quantify, using only calculations and without performing any experiments himself, their exact relationship. He concludes that energy (he called it force) is indestructible and transformable. This mirrors almost perfectly the statement of our modern version of *the first law* (or more precisely *the law of conservation of energy*, or *the first law of thermodynamics*), missing only that energy cannot be created or destroyed.

In 1845, at his own expense, Mayer published a more comprehensive treatment of his ideas, now applying it to physiological problems, which were of his initial personal interest. In 1849, Mayer suffered a mental breakdown and attempted suicide by jumping out the third floor of his house, falling some thirty feet and sustaining a slight, yet permanent disability. The pressure of ridicule, the death of two children, and the impending loss of credit to another scientist, James Prescott Joule, had taken its toll.

Unlike Mayer, James Prescott Joule (1818–1889) performed rigorous and painstaking experiments to test his ideas. In his first experiment, he generated a current in a closed wire, which he immersed in water, thus producing heat[31] (as a result of the resistance or friction between the flowing current and the wire). The current was produced by an electric generator that was driven by a weight falling under the force of gravity through a certain distance. In his next experiment he once again used a falling weight, but this time it was used to turn a paddle wheel immersed in water. The turning of the paddle wheel in the water produced heat as a result of the friction between the water and the paddle wheel.

These two separate experiments convincingly illustrated one fundamental fact: the falling weight provides the work[32] necessary to generate a given amount of heat. Joule was able to quantify the amount of work needed

---

[31] From this study Joule was able to establish that the amount of heat produced was proportional to the amount of current multiplied by the amount of resistance squared, between the wire and the current. This is called Joule heating.

[32] Let's be clear here, we learned earlier that work is accomplished by applying a force over a given distance to, for example, move an object. In this case, gravity is the force moving the object, thus causing it to fall to the ground.

to produce a given amount of heat,[33] thus providing one of the best values of the *mechanical equivalent of heat*. In 1843 he states,

> The quantity of heat capable of increasing the temperature of a pound of water by one degree of Fahrenheit's scale is equal to, and may be converted into, a mechanical force capable of raising 838 lbs. to the [vertical] height of one foot.

Diligent in his efforts,[34] Joule measures the value several times over and gets several values: 820, 814, 795, 760, and many more.

In 1845, the same year Mayer published a more extensive paper based on his original work at his own expense, Joule announced an average value of 817 foot-pounds, and in 1850, after even more measurements, he settled on 772 foot-pounds, which is within 1% of the modern value of 778 foot-pounds.

Nature has set an incredibly high "mechanical price" on the amount of heat one is rewarded with from such efforts as turning a paddle wheel in water. To be sure, consider the amount of heat you get from vigorously stirring your drink – an amount that is totally unnoticeable.[35] And yet in a time without digital thermometers, Joule was able to obtain a value of incredible accuracy.

Neither Joule's published reports nor talks at scientific meetings inspired much interest in his results. In 1847, Joule gave a talk at the Oxford meeting of the British Association for the Advancement of Science. He was asked by

---

[33] The work is simply calculated as the distance the weight falls under the force of gravity, whereas the heat is related to the temperature change of the specific amount of water.

[34] On his honeymoon, Joule supposedly attempted to measure the increase in temperature of the water at the bottom of a waterfall. Indeed, because the water at the bottom has lost (gravitational) potential energy by falling to a lower distance, conservation of energy results in the warmer water at the bottom.

[35] Be grateful that nature has indeed set such high price on the mechanical equivalent of heat. If it were "too low" we would have found ourselves generating scorching amounts of heat from such everyday activities as walking, where heat is generated from the friction between you and the surface you're walking upon.

the chairman to keep his talk brief since it was anticipated that there would be little enthusiasm from the audience. Joule later recalled:

> This I endeavored to do, and discussion not being invited, the communication would have passed without comment if a young man had not risen in the section, and by his intelligent observations created a lively interest in the new theory.

The young man was William Thomson, then only twenty-three years old.

William Thomson (later Lord Kelvin) (1824–1907) quickly recognized the significance of Joule's work. That is not to say that he immediately believed Joule's results. On the contrary, Joule's conclusions were at odds with Thomson's belief in caloric theory and the work of Sadi Carnot (1796–1832), another fan of caloric theory, who insisted that heat was conserved in the generation of mechanical work by heat engines; a loss of heat never occurs in the production of work by a heat engine. This was clearly in conflict with Joule's claim of the equivalence of heat and work, which would mean that the process of mechanical work by a heat engine must result in heat being consumed – not conserved – in the process.

This follows from our discussion of Joule's experiments where he showed that the work done by the falling weight resulted in the production of heat, in accordance with "equivalence": the raising of the weight back to its initial height would require the consumption of that same amount of heat. Indeed, in 1848 Joule told Thomson that he had sought to provide a "proof of the convertibility of heat into [mechanical] power," evidence that Joule himself, unlike Carnot, believed that heat could be transformed into work and that it was fundamental to the theory of the heat engine. Moreover, like Rumford, Joule believed heat to be motion, saying that he had always been "strongly attached to the theory which regards heat as motion amongst the particles of matter."

Eventually Thomson became Joule's biggest proponent, declaring in 1854 in an address to the British Association that Joule's discoveries on heat and work had "led to the greatest reform that physical science has experienced since the days of Newton." In 1866, Joule was awarded the Copley Medal of the Royal Society; Mayer received this honor five years later.

The efforts of Mayer and Joule went a long way at unraveling the nature of heat. Rumford's declaration of heat as a form of "motion" was now more precisely viewed to be the motion of the particles that make up matter. The establishment of the mechanical equivalent of heat once and for all identified heat as just another form of energy and illustrated that energy, at least in this respect, was conserved. However, the understanding of the conservation of energy achieves a new level with the efforts of Hermann von Helmholtz.

The Conservation of All Energy

A physician by training, Hermann von Helmholtz's (1821–1894) interest in energy conservation began with an attempt to prove that the body heat and muscle movement produced by animals was directly related to the energy stored in food. Helmholtz firmly believed that energy was transformed from one form into another while never being created or destroyed. Indeed, Helmholtz coined the phrase "Principle of Conservation of Energy" and proceeded to construct a full mathematical formulation of the conservation of energy as applied to mechanics, heat, electricity, magnetism, chemistry, and astronomy, which is something that Mayer couldn't quite achieve and Joule never tried.

He applied his formulism to a variety of physical phenomena. In particular, he argued that the loss of some of the kinetic energy in inelastic collisions is due to the generation of heat, and the rest is from the deformation of the objects involved. For Helmholtz the deformation was a result of the increase in "tensional force." This formalism is very similar to Johann Bernoulli's argument, that the kinetic energy, or *vis viva*, lost in inelastic collisions became stored by compressing "tiny springs" that he envisioned as making up the object.

Fundamentally both Helmholtz and Bernoulli were correct, and today we understand deformation as changing the potential energy stored within the object. However, Helmholtz's (correct) insight into the generation of heat during an inelastic collision distinguished him from Bernoulli and shed light on the nature of heat beyond its mechanical equivalent of work.

Drawing upon Joule's early papers, Helmholtz went on to apply his conservation principle to thermal and electrical phenomena. He rejected caloric theory and was instead of the view that heat was the result of the constituents of matter being in motion. For Helmholtz, heat and mechanical phenomena were explicitly connected and subject to, like all the other forms of energy he unified under his mathematical paradigm, the first law, which became firmly established around 1850, providing a new unifying framework for physical theory.

Energy the Chameleon

Our initial understanding of energy came from experimental observations such as those made by Galileo in the sixteenth and seventeenth centuries. However, by the late seventeenth century mathematics was a powerful tool in science, as exemplified by Newton's *Principia* published in 1687. Nonetheless, an understanding of energy in its entirety didn't occur until well into the nineteenth century.

Heat was perhaps the biggest obstacle, remaining separate from energy until around 1850, when the first law (the law of conservation of energy, or the first law of thermodynamics) was formalized. Until then, heat was thought to be a sort of fluid that was able to slip in and out of the tiny spaces that were assumed to exist in matter. It went under the name *caloric* and for a long time it was thought to be conserved, separate from the rest of energy. However, as our understanding of matter improved, so did our understanding of heat, and it was finally realized that heat was nothing more than another form of energy. Indeed, we were forced to revise our view on the very nature of matter itself by realizing that heat was nothing more than the motion of its constituent parts.

Today, we recognize many forms of energy: kinetic, potential, chemical, electrical, light energy, nuclear, and heat, to name a few. Energy is truly a chameleon of sorts, able to change from one form to another but never lost; it's always conserved.[36] There's a certain paradox in the fact that we often

---

[36] Further to this point, Einstein taught us that $E = mc^2$, which means something has energy merely because it has mass, m. This equation is often misunderstood. In its

talk about "conserving energy," when, in fact, nature is conserving energy all the time. Of course, what we mean is "don't waste energy." We recognize that, in our hands, energy is not unlimited and it can be used up. Moreover, it means we recognize that not all sources of energy are viable or "useful." That is, not all energy can provide us with a source of work.

This apparent disparity illustrates something very fundamental about energy: while all energy is conserved, not all forms of energy are usable to us. Moreover, when we do use energy for something useful to us, nature requires a certain amount of it to simply be wasted. That is to say, the energy put towards our desired task can never be used in its entirety. Once again, nature is expecting a certain amount of compensation. In fact, these governing principles of energy tie it to another very important quantity: *entropy*.

---

entirety we write: $E = mc^2 = m_0c^2 + KE$, where m_0 is the "rest mass," c is the speed of light, and KE is the kinetic energy.

PART II

Nature's Compensation: Entropy

The [second] law that entropy always increases, holds, I think, the supreme position among the laws of Nature. … if your theory is found to be against the second law of thermodynamics I can give you no hope; there is nothing for it but to collapse in deepest humiliation.

– SIR ARTHUR STANLEY EDDINGTON,
BRITISH ASTROPHYSICIST (1882–1944)

CHAPTER 5

Thoughts of Heat Engines

The Thermodynamic Origins of Entropy

By the 1820s the industrial revolution was in full force, driven both literally and figuratively by the steam engine. A steam engine is a type of *heat engine* that uses steam as its *working fluid*; the steam is the carrier of the heat that is used to produce work. Another version of the heat engine is your car engine. Here, the working fluid is a mixture of gas and air; the combustion of this mixture results in heat production and increased pressure that drives the pistons in the cylinders of the engine, resulting in the motion of your car.

A heat engine requires at least two different temperatures to operate as it converts heat into work. In its simplest form, a heat engine takes a quantity of heat (q_H) from a hot reservoir (*source* at higher temperature T_H), uses a portion to do work (W), and discards a portion (q_C) to a cold reservoir (*sink* at lower temperature T_C), which is usually the surroundings (Figure 5.1).

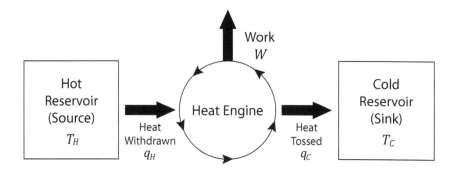

Figure 5.1. Heat (q_H) is withdrawn from the hot reservoir (*source*, at higher temperature T_H). A portion of this heat is converted to work (W) by the heat engine, while a portion (q_C) is tossed to the cold reservoir (*sink*, at lower temperature T_C).

Consider the very simple heat engine of a blown-up balloon and a hair dryer. Placing a small weight on top of the balloon and adding heat with the hair dryer (the source) will cause the air inside the balloon (the working fluid) to expand and lift the weight up as it absorbs a portion of the heat, while the rest of the heat will be lost to the surroundings (the sink). Therefore, our very simple heat engine has done work with a portion of the heat from the hair dryer, while another portion is tossed to the surroundings. Moreover, if we removed the weight and conduct the same experiment, the balloon still does work by expanding against the external pressure. That is, this time it does work by "lifting" the air outside the balloon, whereas before it did this and lifted the weight that was on top of it.

A significant amount of heat is discarded in this process, simply wasted as it passes from the source to the sink. By the 1820s, decades' worth of effort to limit this waste had resulted in an *efficiency* where 6% of the heat from the source was actually used for work, with the remaining 94% lost to the surroundings – pretty poor indeed.[37]

From an economical standpoint, there was much interest in improving the efficiency of the steam engine, or, in other words, maximizing the work obtained from the heat withdrawn from the sink and minimizing (or eliminating) the

---

[37]   Even today, all forms of heat engines throw away a significant portion of heat.

heat loss to the surroundings. This is what Sadi Carnot set out to accomplish, and in doing so, he became a founder of the theory of thermodynamics.

The Heat Engine

Sadi Carnot was the son of Lazare Carnot, one of the most powerful men in France in the late eighteenth and early nineteenth centuries. Lazare was a successful politician and government official, with his brilliance in war logistics and strategy salvaging what would have otherwise been a military disaster in the revolutionary war effort, which earned him the names "the great Carnot" and "the organizer of victory." He was also an accomplished scientist and engineer.

In 1783, Lazare published a memoir in which he approached engineering mechanics from a purely theoretical perspective. In this and later works, he sought to describe the way complicated mechanical machines worked by focusing on the general principles common to them all, rather than getting caught up in all the details specific to individual designs.

In other words, Lazare believed there to be underlying fundamental principles common to all mechanical machines – regardless of the way they were built – that could be used to construct a general theory applicable to them all. He was searching for the "big picture," so to speak. His approach was unique for this time, when others chose instead to draw their conclusions from specific mechanical designs. Sadi Carnot would adopt his father's ingenious approach in the construction of his own theory of heat engines.

In 1823, when Carnot began this task, less than thirty years had passed since Rumford's cannon-boring experiments led him to declare "heat is motion." And although this should have been the end of caloric theory, it and its principle of heat (caloric) conservation were mostly undaunted. Further, a more complete understanding of energy would have to wait for some thirty years for the first law to be established. Thus, it's not surprising that Carnot adopted (the incorrect) caloric theory and that his description of the work done by a heat engine insisted on heat conservation. In addition, he subscribed to another axiom: the impossibility of *perpetual motion*, which had been around a long time in application to mechanical systems. Carnot extended this notion to also include heat engines.

A Divergence on Perpetual Motion of the First Kind

For our purposes, the impossibility of *perpetual motion of the first kind*, simply stated, means you can't create a machine that allows you to obtain more energy than you put into it. Consider this example: you fill the gas tank in your car until it's full, and then drive until you run out of gas. However, instead of your car coming to a screeching halt, it continues to drive down the road because (somehow) you have been able to get more energy out of the engine than was initially supplied by the full tank. Therefore, once you obtained that initial "push" you needed from the full tank of gas, you're able to keep the car running with this "leftover" energy, thus eliminating the need to ever stop at another gas station again.

Clearly, this type of perpetual motion would violate the first law, since obtaining more energy than was put in would result in energy not being conserved. And while a perpetual motion machine as described may seem absurd to you, be assured that people today continue to search for such machines. In fact, the United States Patent and Trademark Office's (USPTO) official policy regarding perpetual motion machines is to refuse granting a patent without a working model. This is clearly stated:

> With the exception of cases involving perpetual motion, a model is not ordinarily required by the Office to demonstrate the operability of a device. If operability of a device is questioned, the applicant must establish it to the satisfaction of the examiner, but he or she may choose his or her own way of so doing.

To be sure, the USPTO will file the patent application as this is done by a patent clerk, but the actual issuing of a patent, done by a patent examiner,[38] will most likely not happen. Nonetheless, the patent office did issue a patent in 1979 (U.S. Patent 4,151,431) for a device appearing to be a perpetual motion machine, which illustrates the point that just because an invention is issued a patent does not mean it will actually work.

---

[38] Patent examiners have technical degrees, and some even having PhD qualifications.

After understanding the first law, you may wonder how such perpetual motion can even be a topic of debate, since as described it's clearly a violation of the first law (hence perpetual motion of the "first kind"). However, the first law is not something that has a rigorous mathematical proof associated with it; it's not something that one can sit down and derive by using mathematics and known physics. It's simply a statement – a very strong one indeed – of something we believe to be true based on all the experimental evidence that has been compiled to date. We simply haven't found a single violation of the first law thus far, and therefore we hold it to be true.

Carnot's Reversible Heat Engine

Carnot developed a mathematical model of a heat engine that consisted of the working fluid; the hot reservoir, where heat was withdrawn by the working fluid; and the cold reservoir, where a portion of that heat was dumped by the working fluid. That's it. The specific details of the heat engine were deliberately excluded – the type of work being done, the mechanical design (i.e., the moving parts), and the type of working fluid used were all intentionally ignored.

Recall that in practice, upon absorbing heat from the source, the working fluid of a heat engine[39] is able to produce work by expanding

---

[39] To be sure, not all heat engines work in this way. A very simple heat engine is the "drinking bird." As it dips its beak into the glass of water in front of it, the drinking bird absorbs water onto its beak and then returns to its upright position. From here, the water begins to evaporate from the tip of its beak, causing it to cool – a process known as *evaporative cooling*. As a result of this cooling, the top portion of the bird becomes colder (cold reservoir) than the bottom portion (hot reservoir), resulting in a temperature difference. The effect of this temperature difference is that the liquid inside the bird rises to the top until finally the bird tips over into the glass of water to take another "sip," and the whole process begins again. Moreover, the temperature difference also results in a pressure difference, such that the pressure at the top is lower than the pressure at the bottom. The liquid (working fluid) inside plays a crucial role in the whole process as well. The liquid inside is very volatile, which means at room temperature a significant portion of the molecules are in the vapor phase in addition to those in the liquid phase. In general, all liquids exist such that some molecules are in the liquid phase with others in the vapor phase,

in a confined space, which in turn drives moving parts into motion. For example, in your car, the pistons are enclosed in the confined space of the cylinders, and the heat comes from the combustion of the air-gas mixture (the working fluid) within these cylinders that leads to an increase of pressure on the pistons, resulting in the work needed to create the motion of your car driving down the road.

Carnot ignored all these little details in his attempt at deriving a general theory of heat engines, requiring only that his heat engine be *reversible*. In doing so he not only accomplished his goals, but also introduced a whole new mathematical model into thermodynamics, which is still in use today.

Let's talk about what it means for a process to be reversible (or to move a system *reversibly*). If a process is reversible, it means that the system involved in the process can be returned, or reversed, to its prior state *exactly* as it was before the process occurred. That's really all there's to it.

So what's the big the deal? Actually, no processes in nature[40] – the things we're interested in really understanding – are reversible; rather, they are

---

which gives rise to the *vapor pressure* of a liquid. However, volatile liquids simply have "more" of their molecules in the vapor phase compared to less volatile liquids at the same temperature. Physically this is a reflection of the attractive interactions between the molecules of the liquid; the more attractive the interactions are, the more molecules that remain in liquid phase and the less in the vapor phase, and vice versa. This is a key point because a more volatile liquid will respond more dramatically than a less volatile one to a pressure difference, like the one inside the drinking bird, with more molecules leaving the liquid phase by evaporating into the vapor phase. It's this evaporation, along with a bit of *capillary action,* that causes the liquid to rise upward.

[40]   With the exception of systems and processes in equilibrium, the majority of systems and processes that occur in nature are irreversible. Such equilibrium systems include those undergoing a phase transition. For example, when water is at its freezing/melting point, or when it's at its boiling/condensation point, the system is in equilibrium and is therefore reversible. Other examples include chemical reactions that are in equilibrium. This means the forward and the reverse reactions are proceeding at the same time. However, it doesn't mean they are occurring at exactly the same rate, and in fact, they are usually not. This latter consideration is related to the chemical *kinetics* of the reaction, whereas the former consideration has to do with the *thermodynamics* of the reaction.

irreversible. A well-known irreversible process is that of an egg falling to the floor and "splatting." Clearly, it can never be returned to its previous "un-splatted" state.[41] Moreover, much understanding of the processes in nature comes from mathematical models. Unfortunately, it's the reversible processes that are easier to handle mathematically (often leading to insightful results), not the irreversible ones. So, if possible, it's advantageous to approximate our irreversible process via a corresponding reversible one.

Let's consider the following: two blocks of exactly the same weight are glued to a teeter-totter at exactly the same distance from the center pivot point (what a strange playground indeed). As such, one block's weight is just enough to balance the weight of the other block regardless of the orientation of the teeter-totter; the force of their respective weights perfectly balances at every position throughout. We find one block all the way down on the ground, and the other block raised all the way up.

Now let's say we want to move the block on top all the way down, such that at any time along the way the movement is reversible. Well, this seems easy enough. After all, reversible in this case means simply that we can return the teeter-totter and the blocks back to their prior position any time throughout the movement. So, we begin to move the teeter-totter downward, and no sooner do we get started than we hear a squeaking noise – oh yeah, the pivot point is rusty and there's a lot of friction; this is going to be a problem.

The challenge here is to move the teeter-totter reversibly, and that annoying squeaking noise is reminding us of friction, and that heat is being produced as a result of it. Sure, it probably isn't that much heat; it's probably barely noticeable to the touch. Nonetheless, this means we have failed to move the teeter-totter reversibly. This is clear, since if we simply move it back to its previous position, everything will be the same – except for the heat; we can't simply pull the heat out of the air and put it back into the pivot point, after all. So how do we make this process reversible?

---

[41] You might be imagining a glass that has fallen off the counter and broken into a few pieces. It's tempting to think that by merely gluing all the pieces carefully back together we have managed to reverse the glass to its unbroken state. Certainly we have not; at best we have only approximated the unbroken state it existed in before.

Let's begin again, but this time we apply only a small force and move the teeter-totter over a small distance. Since the blocks are of the same weight, only enough force to overcome the bit of friction at the pivot point is required to unbalance them. Using a small downward force applied to the block on top, we move them a very small distance, and then stop. The block on top has now moved very slightly downward while the block on the ground has moved very slightly upward. This time there's no squeaking noise and thus no heat is produced (we assume for the sake of our argument); we have moved the teeter-totter reversibly. Further, once we stop providing the additional force, the blocks also stop, remaining fixed and perfectly balanced at their new positions; in other words, they have reached an *equilibrium* position.

Now, without stopping, we again apply a tiny force to move the block on top ever so slowly through a series of tiny steps until we reach the ground. Although we didn't stop along the way as before, we are assured that the teeter-totter moved through a series of equilibrium steps that were reversible and without heat production. Our assurance comes from the fact that we used small forces that moved the teeter-totter very slowly via small steps. As you can imagine, moving the teeter-totter the total distance took a long time because of this approach – much longer than it would have taken had we simply moved it in an irreversible fashion. Unfortunately, that's what it took to accomplish reversibility in this case: small forces, small steps, and a large amount of time. In fact, that's pretty much what it takes to move any system reversibly. Why?

Well, simply put: because that's what it takes to not disturb a system so much that it can't be put back the way it was before. Is it any wonder that systems in nature, or everyday life, are not reversible? The concept of reversibility makes for a great mathematical model, whereby we gain insight into everyday occurrences, even if the universe doesn't exactly run this way.

For a heat engine, reversibility applies not only to *mechanical reversibility*, as just discussed for the teeter-totter, but also to *thermal reversibility*. In our teeter-totter example, we were concerned with moving the block on top a certain downward distance. In a heat engine, there's a similar mechanical process, but there's also a temperature difference (*temperature gradient*), like the downward distance over which you want to "move" heat. Thermal reversibility is accomplished by having the heat move, or flow, from the hot reservoir to the cold reservoir in "tiny steps."

This is accomplished in a similar fashion as before, except this time we want to apply only the smallest of *thermal forces* to accomplish the flow of heat. Specifically, we want the heat to flow from the heat source to the cold sink, moving through regions where the drop in temperature is minuscule. In practice, this means that one would want hot and cold portions of the heat engine (mechanical parts and working fluid) that are in contact with each other to differ very little in temperature.

For sure, reversibility is an idealization. It's the best-case scenario we strive for in design but can never achieve. It's simply impossible to move a system in such small steps, either literally or practically, as it would take an extremely long time – essentially forever – to accomplish one's desired task.

Moreover, whether it's mechanical friction, such as that between a real teeter-totter and its pivot point, or thermal friction, which exists whenever parts of different temperatures are transferring heat (like in your car engine), we simply can't eliminate it altogether. Nonetheless, reversibility provides us with a powerful mathematical model, while setting an unobtainable upper bound (a sort of gold standard) for all real systems. For Carnot it provided profound insight into the nature of real heat engines and pointed the way towards a new physical quantity.

From Hot to Cold

Carnot realized that heat flows from hot to cold, and that a heat engine allows one to obtain work from this flow. He imagined this temperature difference as similar to the height difference needed to run a water engine. An example of a water engine would be a water wheel positioned at the bottom of a waterfall. The water falling from the top to the bottom[42] turns the water wheel when it hits it on its descent, and this motion can be

---

[42] The energy to do work comes from the potential energy due to the height difference.

harnessed to do work. A water wheel is the most efficient[43] when every single drop of water falling from the top hits the water wheel and drives it on its way to the bottom; water falling past the water wheel does not contribute to this motion and therefore lowers the efficiency.

Likewise, Carnot envisioned something similar occurring in heat engines. Moreover, he assumed that it's impossible to extract work from heat without having a temperature difference; there must be a hot reservoir (source) and a cold reservoir (sink) for heat to flow and drive a heat engine, just as there must be a height difference (from high to low) for water to flow to drive a water engine.

He was also convinced that heat was necessarily discarded in this effort. As in a water engine, where the water falls from the top to join the rest of the water below, Carnot concluded, so must heat in a heat engine "fall" from high temperature to low temperature, finally being tossed into the cold reservoir.

In a water engine, all the water falling from the top ends up at the bottom, and in this way the water is conserved (except for some evaporation, of course). Carnot, as a believer in caloric theory and adhering strongly to the water engine analogy, believed that the heat in a heat engine was conserved as well and that all the heat from the hot reservoir would end up in the cold reservoir in the process of running a heat engine.

After the discovery of the first law some thirty years later, it became clear that it's not heat that is conserved, but rather energy as a whole. Therefore, the amount of heat initially withdrawn from the hot reservoir equals the amount of heat tossed to the cold reservoir plus the work done by the heat engine.

It was Carnot's mathematical construction of a reversible heat engine that led him to his most important conclusions. A thought experiment will help us appreciate his model. Suppose we have two of Carnot's reversible heat engines (Figure 5.2). We take the heat engines; let's call them Engine 1 and Engine 2 and connect them to the same hot and cold reservoirs. Now,

---

[43] There's some loss of potential energy due to heating when the water hits the waterwheel, which lowers the efficiency.

we imagine getting two different work outputs from each. For clarity let's call these amounts W_1 and W_2 and respectively, where W_1 is greater than W_2 In other words, the efficiency of Engine 1 is greater than that of Engine 2.

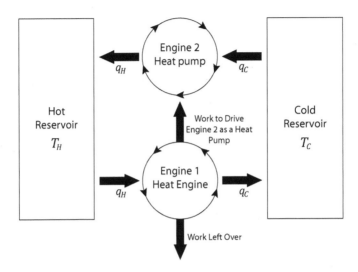

Figure 5.2. We imagine Engine 1 withdraws an initial amount of heat (q_H) from the hot reservoir. A certain portion of this (q_C) is tossed to the cold reservoir, while the rest (it's reversible, so there are no heat losses due to mechanical or thermal friction) is used for work. Some of this work is used to drive Engine 2 as a heat pump that withdraws from the cold reservoir the same amount of heat (q_C that was tossed by Engine 1) back out, adds a portion of heat of its own, and is now able to toss the full initial amount of heat (q_H withdrawn by Engine 1) back into the hot reservoir. We further imagine that Engine 1 can do all this and still have an amount of work left over to do something else.

The key is the reversibility of the heat engines, which means there's no mechanical or thermal friction to overcome. Therefore, it only takes a minuscule amount of work – in addition to the work output – to turn either into a *heat pump* that takes heat from the cold reservoir and dumps it to the hot reservoir. This is similar to our reversible teeter-totter, in which case at any point along the way we could have easily reversed directions by applying only the smallest of forces. A familiar heat pump is the refrigerator, which cools by taking the heat from the inside and discarding it to the outside.[44] Indeed, a heat pump is similar to a *water pump*, which is simply a water engine that has been made to run in reverse, causing it to move water from a low height to a high height.

Since Engine 1 has the larger amount of work output (W_1), we use it to drive Engine 2 in reverse, turning it into a heat pump. In this way, we can now create the following cycle: Engine 1 withdraws an initial amount of heat (q_H) from the hot reservoir, tosses a portion of it (q_C) to the cold reservoir, and does work that drives Engine 2 to withdraw the same amount of heat (q_C that was tossed by Engine 1) back out of the cold reservoir, and to add a portion of heat of its own, such that it can now toss the full initial amount of heat (q_H withdrawn by Engine 1) back into the hot reservoir. Moreover, we can imagine that Engine 1 can do all this and still have an amount of work left over to do something else. Finally, this all happens reversibly, so there are no heat losses due to mechanical or thermal friction.

If something seems wrong to you, you're not alone; Carnot thought so too. In short, the net result from this cycle is that Engine 1 is able to get work out without having to use energy in the first place. It does this by having Engine 2 (the trusty heat pump) simply put back the energy it had to initially withdraw from the hot reservoir (q_H). And even after it does all this, it still has an amount of work left over to do something else. Moreover, since the hot reservoir never runs out, this cycle continues forever! Although different in details, this scenario is the same as the perpetual motion of the first kind discussed earlier, where we imagined one's car running forever on a single tank of gas.[45]

---

[44] Another is an air conditioner.

[45] Perhaps things could be resolved by making the cycle such that Engine 1 can only perform the work required to run Engine 2 as a heat pump and nothing else. After all, the net result then is that heat is simply taken from the hot reservoir and then placed back into it. In fact, we can simply remove the cold reservoir altogether. What could really be the harm here? Well, now we have the scenario where at a single temperature (T_H), heat is actually flowing back and forth between the heat engine and heat pump. In other words, we've created a general physical mechanism whereby a system at *thermal equilibrium* can still experience heat flow. However, as

In adherence to his main tenets for his general theory, Carnot concluded that the work output (or efficiencies) of any two reversible heat engines operating between the same temperature difference must be the same, thus eliminating the construction of our would-be perpetual motion machine. More precisely, *Carnot's Theorem* says that the work output (and thus efficiency) of any reversible (ideal) heat engine depends only on the temperatures of the hot reservoir and the cold reservoir. Although I won't derive it here,[46] in terms of efficiency we can write this statement mathematically as:

$$Efficiency = 1 - T_C / T_H$$

where T_C and T_H are the temperatures of the cold and hot reservoirs respectively, and a value of 1 represents 100% efficiency. Therefore, the amount of heat that a reversible heat engine can withdraw and the amount of work it can perform are solely determined by the temperature difference. In contrast, a real heat engine would suffer a loss on both accounts because of its physical design, which we'll discuss in a moment. In other words, a reversible heat engine represents the maximum amount of work and efficiency any heat engine can achieve for the temperature difference.[47]

---

discussed, heat can't flow without a temperature difference, and thermal equilibrium means heat doesn't flow at all. Moreover, such a device would violate *the second law* (as we'll better understand later); it would be *perpetual motion of the second kind*. So, we are forced to reject this as a physical possibility.

[46]   This formula is easily derived by using an ideal gas as your working fluid and then running it through a four-cycle process: 1) isothermal (at constant temperature) expansion at the temperature of the hot reservoir; 2) adiabatic (without heat lost or absorbed) expansion resulting in cooling to the temperature of the cold reservoir; 3) isothermal compression at the temperature of the cold reservoir; and 4) adiabatic compression, which results in heating the gas back up to the temperature of the hot reservoir, thus returning the gas to its original state, ready to start the cycle all over again. Recall that by design, Carnot's reversible heat engine doesn't depend on the specific design. Therefore, the use of an ideal gas as the working fluid is merely a mathematical convenience and doesn't affect the resulting formula.

[47]   The efficiency is defined as the magnitude of work divided by heat input from the hot reservoir (the source).

Indeed, moving a system reversibly is the best-case scenario, and the resulting work output and efficiency of a reversible heat engine provide an upper bound that no real heat engine can surpass. Nonetheless, according to our equation we still can't get 100% efficiency, even for the best-case scenario of the reversible heat engine.[48] This is because a temperature difference (between T_C and T_H) is still needed to get the required heat flow.[49] This is a restatement of the fact, as mentioned earlier, that one must discard a portion of heat to the cold reservoir, and therefore not all of the heat withdrawn from the hot reservoir can be used for work.

In other words, if you want to use a heat engine to do work for you, nature again requires its compensation. This is not unlike before when (in Part I) we learned that the price for using a simple machine, like the inclined plane, also required a given amount of compensation. Whether it's an inclined plane or heat engine, nature isn't giving anything away for free. Realize that we arrived at this result without considering the details of the working fluid or the specific design of the heat engine (other than the fact that it's reversible). Consequently, for a reversible heat engine, the specific design details are ignored. To be sure, the working fluid, the mechanical parts, or even the materials you use to build your heat engine are neglected altogether; all that is important is that it runs reversibly.

Carnot's Theorem teaches us about real heat engines (like your car engine) that run irreversibly. As mentioned before, no real heat engine will ever be more efficient than its reversible counterpart. Nonetheless, we still attempt to make our real machines "more reversible." This means minimizing sources of mechanical friction between the moving parts and avoiding thermal friction due to large temperature differences between the moving parts and working fluid of the heat engine.

---

[48] The combination of the first law, which requires energy to be conserved, and the fact that no heat engine can obtain 100% efficiency is a testament to the physical consequence of having to toss a portion of heat from the hot reservoir to the cold reservoir when running a heat engine.

[49] In fact, notice that if $T_H = T_C$, the efficiency is zero. This means you can't get work without a temperature difference, which is essentially Thomson's statement of *the second law*.

In practice, the mechanical design, the materials used, and the working fluid all play a role in the efficiency of a real heat engine. For the working fluid, such physical properties as the time it takes for heat to flow through the working fluid (the thermal conductivity), the melting point, the boiling point, the surface tension, and the vapor pressure will all affect the efficiency of a real heat engine. As indicated from Carnot's Theorem, improvements in efficiency result from a larger temperature difference between the hot and cold reservoirs. Since the cold reservoir is the temperature of the surroundings, which can't be controlled (e.g., you can't control the outdoor temperature your car runs at),[50] this difference is accomplished by making the hot reservoir hotter. Carnot's ingenious model provides much to engineers in the way of real heat engine design.

In 1824, a year after his father's death, Carnot wrote *Reflections on the Motive Power of Fire*, in which he emphasized, in a fashion similar to his father's work, his general theory being applicable to all types of heat engines regardless of the details of their design. The memoir was published by a leading scientific publisher, received only one – albeit enthusiastic – review, and a decade later was cited in an important journal. The tome then spent the next twenty years in obscurity.[51]

In June 1832, Carnot contracted scarlet fever. When he was feeling momentarily better he wrote to a friend:

---

[50] Something to think about next time you drive your car: it runs more efficiently when it's colder outside.

[51] While it's not totally clear why Carnot's memoir didn't leave a greater impact on its intended audience of physicists and engineers, it might have been the style in which it was written. Carnot wrote his memoir in a rather popularized style, rarely using mathematical equations, and when he did he included them modestly in footnotes. However, for engineers the work was still too theoretical; and scientists, who were accustomed to a theoretical approach, didn't take the work seriously either, as they found it lacking (necessary) mathematical rigor. In the end, Carnot was unable to connect with either group and ended up missing his audience altogether. Carnot did nothing to promote *Reflections*. He devoted his time instead to an intense research regimen and never published again.

> I have been sick for a long time, and in a very wearisome way. I have had an inflammation of the lungs, followed by scarlet-fever. (Perhaps you know what this horrible disease is.) I had to remain twelve days in bed, without sleep or food, without any occupation ....

The scarlet fever eventually spread to his brain, and then in August he contracted cholera and died within hours. He was only thirty-six years old. As was standard with cholera victims, his clothes, his personal effects, and almost all his papers were burned. From the surviving papers we see Carnot beginning to abandon caloric theory as he realizes its inherent challenges in the face of Rumford's prior work (discussed in Part I).

In 1834, Émile Clapeyron (1799–1864), a former classmate of Carnot's, published a paper in the *Journal de l'École Polytechnique*. The paper, reformulating Carnot's work using clear, concise mathematics and a new graphical presentation for Carnot's reversible heat engine (still taught today to every chemistry major in a good physical chemistry class), finally brought Carnot's work to the attention of engineers, chemists, and physicists. Its German and English translations provided the only link to Carnot's work for nearly ten years. It was Clapeyron's version of Carnot's work that was most widely read among contemporary scientists, including William Thomson[52] as a recent graduate from Cambridge and a young German student at the University of Halle, Rudolf Clausius (1822–1888).

Carnot attended the École Polytechnique, where he was surrounded by and sat under renowned physicists, chemists, and mathematicians. Nonetheless, other than during his academic studies, Carnot was never a member of this distinguished group, conducting his important work as an outsider. Regardless, Carnot's contributions to the foundations of thermodynamics are undeniable, and they firmly establish him as both a pioneer and one of the greats in this field.

---

[52] Evidently Thomson searched the Paris bookshops for a copy of Carnot's original work without success.

CHAPTER 6

Dissipation

The Relationship Between Heat and Work

The establishment of the first law, around 1850, provided a clear statement describing the ability of energy to be interchanged among its many different forms while never being created or destroyed. In this way, it's always conserved. The journey leading to this concise statement was driven by the fundamental belief that nature itself involves the unification of seemingly disparate entities[53] and resulted from the tremendous effort of many scientists over many centuries. Along this seemingly wayward journey, several other truths were revealed.

Heat was now believed by most to be the motion of the fundamental particles that constitute all matter (which we would later call atoms and molecules). This was a tremendous step forward not only in our understanding of heat, but also in our developing picture of the essence of matter; in our quest to understand energy and heat we also evolved our picture of

---

[53] The unification of certain things in nature has always been an underlying theme in science. In physics, there has long been the desire to unify the four fundamental interactions or forces: electromagnetism, strong nuclear force, weak nuclear force, and gravity. The unification of electric and magnetic forces led to electromagnetism, and subsequent unification of electromagnetism with the weak force resulted in the electroweak force.

matter. Joule's experimental diligence had rooted the mechanical equivalent of heat in most people's minds, as he had undoubtedly shown that a given amount of work would produce an equivalent amount of heat. Moreover, Joule argued that the same amount of heat would then generate an equivalent amount of work; heat can be used to create work. The days of describing heat as a weightless "imponderable fluid" that was always conserved, according to caloric theory, were nearing the end.

The Problem with Heat and Work

William Thomson's talents spanned many areas: mathematics, physics, engineering, teaching, and politics. He was also a natural problem solver, largely attributable to his exceptional talent in mathematics. By the time of his graduation from Cambridge in 1845, Thomson had already published twelve papers on topics in pure and applied mathematics. His tradition of hard work and publications would continue throughout his lifetime. Every year between 1841 and 1908, he published at least two papers, sometimes as many as twenty-five. In all, he wrote 661 papers and held patents on sixty-nine inventions.

While only sixteen, Thomson read *Analytical Theory of Heat* by Joseph Fourier (1768–1830), published in 1822. In our study of the Carnot heat engine, we talked about heat flowing from a hot reservoir to a cold reservoir. In Fourier's theory, heat flows or is conducted through an object (in reversibly small steps, in fact) due to temperature differences (*gradients*) that exist in the object; in a sense, the object has regions that contain tiny hot and cold reservoirs.

A striking feature of Fourier's theory is that it completely avoids speculation about the physical nature of heat. Rather, Fourier focuses on only the behavior of heat. Thus, the possible causes of heat, whether due to caloric or the motion of tiny particles that make up matter, were of no concern to Fourier's development of his theory.[54]

---

[54] A theory such as Fourier's, where the behavior of the thing in question (e.g., heat) is understood but less so are the underpinnings of why this behavior manifest itself, is known as a phenomenological theory.

Fourier's resulting mathematical equation (specifically, *differential equation*) successfully accomplished exactly what he set out to do: correctly describe the observable behavior of heat, which must be governed by fundamental laws of nature, without getting too hung up on the exact details giving rise to this behavior. The beauty of this approach[55] is that it allows one to move forward even though all the pieces of the puzzle are not fully in place yet, so to speak. In his own words, Fourier describes his goal:

> Primary causes are unknown to us; but are subject to simple and constant laws, which may be discovered by observation, the study of them being the object of natural philosophy.

> Heat, like gravity, penetrates every substance of the universe, its rays occupy all parts of space. The object of our work is to set forth the mathematical laws which this element obeys. The theory of heat will hereafter form one of the most important branches of general physics.

Clearly, Fourier's theory and Carnot's theory of heat are quite different. Carnot taught us to imagine heat (or "caloric," as he called it) as falling, like a waterfall, from high to low temperature and, in this way, one is able to do work. Fourier simply said that regardless of what heat actually is, its nature is such that it spreads as the result of tiny temperature differences throughout an object. Fourier says nothing of being able to use heat to obtain work, as Carnot had, and in fact is essentially saying one need not accomplish any work in this process.

This was a bit unsettling to Thomson. On the one hand, you have Carnot's theory with its reversible heat engine illustrating that one can get a maximum amount of work – in this case, when heat falls from high to low temperature. But on the other hand, you have Fourier's theory saying

---

[55] We are already familiar with this approach from our study of Carnot's reversible heat engine. Recall that Carnot eliminated major details from his idealization of the heat engine such as the working fluid, the mechanical design, and the materials used in the construction. This type approach is often used in the physical sciences.

that heat can spread, once again, from hot to cold regions and do no work at all. To be sure, each of these theories was consistent with experimental observations, thus both appeared to be valid. However, in Thomson's mind this difference was blatant, and surely something must be lost when one went from a Carnot description to a Fourier description. Nonetheless, Thomson was unable to find any inconsistencies between the two theories. And if this confusion wasn't enough for Thomson, it was about to get worse.

In 1847, Thomson met Joule at a meeting of the British Association for the Advancement of Science. Joule talked about his studies on the mechanical equivalent of heat, clearly showing that a given amount of work will produce a given amount of heat (discussed in Part I). Moreover, Joule was convinced from his studies that this conversion could happen in the other direction: a given amount of heat would be able to produce a given amount of work as in a heat engine. Now, much to his dismay, Thomson had three theories to reconcile: Carnot's reversible heat engine, Fourier's "work-free" conduction of heat, and Joule's conversion of heat into work ("heat equivalent of work").

Thomson was disturbed by Joule's claim that heat was actually converted into work in a heat engine. Recall Carnot's claim that one can get work from a heat engine as a result of heat "falling." In Carnot's mind, no heat is used up to produce work; you begin with a given amount of heat, it creates work, and then that exact same amount of heat is tossed to the surroundings; there's no conversion of heat into work. Carnot was convinced that in this way you produced work, and – being a strong proponent of caloric theory – that heat was conserved. While Thomson was now comfortable with abandoning some of caloric theory – the part that envisioned heat as a collection of particles moving as an imponderable fluid – he wasn't ready to renounce caloric theory altogether. Carnot had adopted the idea of heat conservation from caloric theory, and Thomson saw no reason to immediately abandon this tenet either.

Thomson felt Joule's conclusions were also at odds with Fourier's theory of heat conduction. Joule's experiments had demonstrated that work could be converted into heat. Further, Joule stated that his results also implied that heat could be converted into work. However, Fourier's theory correctly described the conduction of heat as it freely passed through an object. What Thomson wondered, however, is what becomes of the "mechanical effect" that Joule saw in his experiments. In Thomson's mind, this effect appeared to be lost.

In 1849, Thomson published one of his first papers on the theory of heat and pointed out that no mechanical effect was observed when heat freely passes (is conducted) through a solid object:

> When "thermal agency" is thus spent in conducting heat through a solid, what becomes of the mechanical effect which it might produce? Nothing can be lost in the operations of nature – no energy[56] can be destroyed. What effect then is produced in place of the mechanical effect which is lost? A perfect theory of heat imperatively demands an answer to this question; yet no answer can be given in the present state of science.

While Thomson was clearly identifying fundamental concerns with the current state of heat and work, he wasn't yet providing any answers. Even looking to the experimental evidence didn't provide a path of reconciliation. However, the answers were just around the corner.

Work from Heat: Joule Versus Carnot

In 1847, Thomson couldn't figure out how to reconcile Carnot's reversible heat engine, which demanded both the conservation of heat (heat withdrawn from the hot reservoir always equals heat tossed out to the cold reservoir) and the generation of work, with Joule's mechanical equivalent of heat, which insisted on the consumption of heat in order to produce work in a heat engine. Nonetheless, Joule and Carnot agreed that a heat engine could utilize heat to produce work. However, they disagreed on what

---

[56] Thomson's use of the word *energy* was a first step towards its current meaning today. This was significant because *energy* had a long history of being used very ambiguously (and incorrectly), often interchangeably with the word *force*.

became of the heat in this process. Carnot believed heat to be conserved, an axiom he borrowed from the then-popular caloric theory, while Joule imagined essentially the opposite, namely that heat was converted to work and thus was lost, not conserved, in the process. Like Carnot, Thomson simply was not ready, at the time, to abandon the concept of heat being conserved.

In 1824, Carnot had led the way with his introduction of the (idealized) reversible heat engine and all the insight that followed from it. To be sure, Carnot was a pioneer who had devised an ingenious mathematical model that resulted in amazing insight, albeit ever so slightly tainted by the fatally flawed use of caloric theory. However, by 1850 the landscape of thermodynamics was changing, and caloric theory was about to see its last days, as the first law – that energy as a whole, rather than heat alone, is conserved – began to take its rightful place thanks to the works of Mayer, Joule, and Helmholtz. In part, this opened the door for Rudolf Clausius to correct the disparate views on heat and work between Carnot and Joule.

Clausius was born in 1822 in Köslin, Prussia (now Koszalin, Poland), and was the youngest of eighteen children. He received his early education at a small private school where his father was the principal. Although initially attracted to history, he received his PhD in mathematical physics from the University of Halle in 1847. He would become interested in electricity and magnetism, even developing a theory on charged atoms in solution (electrolytes), sometimes referred to as the Williamson-Clausius hypothesis (which seems unjustified given that Williamson never considered these types of solutions). He also contributed to *kinetic theory*, introducing the notion of *mean free path*, which is the average distance a particle (molecule or atom) in a fluid freely moves around until it "bumps" into another particle. However, his most significant works occurred in thermodynamics, where he studied the theoretical aspects of the mechanical equivalent of heat, the first law, and, without any doubt his most significant work, the discovery of entropy.[57]

---

[57] Most of Clausius' work was done before 1870, which was likely due to two major life events. In 1870, he was severely wounded in the knee while serving in an ambulance corps in the Franco-Prussian War and suffered lifelong pain from this injury.

In 1850, Clausius published a memoir that finally reconciled the work of Joule and Carnot. Unable to find Carnot's original work, Clausius, like Thomson, learned of Carnot's theory from Clapeyron's article, published two years after Carnot's death. Both Carnot and Clapeyron had been misguided by caloric theory's insistence on the indestructibility of heat (its conservation), which forbids its consumption by a heat engine in the production of work. For them, work was created as heat "fell" from high to low temperature without ever changing in amount.

Clausius saw heat as having two fundamental processes available to it: *conduction* and *conversion*. In the case of a heat engine, Clausius assumed that part of the heat taken from the hot reservoir was converted to work, while the remainder (not used for work) was freely conducted from the hot reservoir to the cold reservoir; this portion of heat becomes the heat engine's output.

Clausius maintained that Joule's experiments had firmly established the equivalence between heat and work. Moreover, he asserted that the "fundamental principle" of Carnot's theory was that heat naturally passes from a high temperature to a low temperature, and in this way work can be done. Therefore, rather than choosing between the two theories of Carnot and Joule, Clausius simply combined them into a more complete theory. In Clausius' own words:

> for it is quite possible that in the production of work both may take place at the same time; a certain portion of heat may be consumed, and a further portion transmitted from a warm body to a cold one; and both portions may stand in a certain definite relation to the quantity of work produced.

By 1850, Thomson was also starting to come to the same realization as Clausius. He finally let go of caloric theory altogether and its major tenet of heat conservation, and in doing so he was now free to accept that heat

---

Moreover, his wife died tragically in childbirth and he assumed the responsibility of raising six young children. Late in his life, when he was in his sixties, he married again.

could indeed be converted to work as Joule had stated. This allowed him, as it did for Clausius, to reconcile the theories of Carnot and Joule in the same fashion. Much to his relief, Thomson found that the elimination of heat conservation from Carnot's theory still preserved the mathematical equations he had derived from it originally. In 1851, almost a year after Clausius' work, Thomson published *On the Dynamical Theory of Heat*, therein discussing his harmonization of the theories of Carnot and Joule but acknowledging Clausius' priority.

Energy, Work, and Heat

Thomson's major contribution in *On the Dynamical Theory of Heat* may have been his perspective on the energy of a system. He shifted the emphasis away from heat and work, a focus that had been originated by Carnot and carried on by Joule and Clausius, and placed it instead on energy.

Thomson defined energy as an inherent property of the system; all systems have energy as a fundamental property. Moreover, he stated that a system's energy could only change through interactions with its *surroundings*. Therefore, if a system is completely isolated from its surroundings, its energy can't change but rather is conserved.

Notice that this statement makes no reference to what the system contains; nor does it mention any details of what's actually occurring inside – it doesn't matter. As long as the system is isolated from its surroundings, nature has provided no means by which it can change its energy; the energy will forever be conserved. This is a very powerful statement indeed. It freed Thomson from having to speculate on the nature of matter within the system.[58]

We learned about a variety of systems in Part I, in particular, those studied by Galileo, like the ball rolling down the inclined plane. At that

---

[58]   As we learned in Part I, the nature of matter confounded much of the early understanding of energy, in particular of heat. When Thomson wrote *On the Dynamical Theory of Heat* in 1850, many, such as Joule, Mayer, Helmholtz, and Clausius, were beginning to accept matter as being made of constituent particles (which we today call atoms), but no one was quite ready to commit to it altogether in terms of providing a theoretical formulation.

time, we didn't talk about what actually constituted the system and the surroundings. So let's clarify it now.

By *isolated* we mean nothing can leave or enter the system: no matter, no heat, and no work can be done *by* the system or *on* the system. Imagine an inclined plane with a ball poised at the top, ready to roll down as soon as a slight push is given. Now, the inclined plane, the ball, and yours truly are all placed in a building. Once everything is inside, the door is locked from the outside. Our system consists of everything inside the building. Clearly, no matter can leave or enter the building.[59] Further, imagine the walls have been thoroughly insulated so that no heat can pass in or out. So, we have ensured no matter or heat can enter or leave our system.

What about work? We learned in Part I that work is accomplished by applying a force to an object to move it over a certain distance.[60] If an *external force* were supplied from outside our system *by the surroundings* then possibly it could do work *on* our system. For example, if the ball sitting at the top of the incline were made of a magnetic material such as iron, a strong magnetic field from outside the system could be applied that would nudge the ball, causing it to roll down the inclined plane. In this way, work would be done *on* the system by the surroundings.

Moreover, in a similar scenario you can imagine a force from the system (inside the building) causing work to be done on the surroundings; now we would say work has been done *by* the system on the surroundings. But once again, we have eliminated such possibilities and have secured our system (inclined plane, ball that is ready to roll, and me) within the confines of the building, totally isolated from its surroundings. According to Thomson, the energy of the system should now be conserved regardless of what happens inside. Let's test this idea out.

---

[59]   For completeness, imagine air cannot leave or enter the building; I'm on the clock with my little experiment here.

[60]   Here we have talked about work in terms of a heat engine converting heat to work. Realize that our original definition of work still applies. In the case of a heat engine, the working fluid absorbs heat that causes it to expand. As a result of this expansion, the working fluid ends up pushing on an object, thus causing it to move over a certain distance; the working fluid applies a force to move the object and thus does work.

I proceed to push the ball, causing it to roll down the inclined plane. My push supplied the ball with an amount of energy, which results in a loss of energy for me. The ball rolling from its initial height down to a lower height changes all its potential energy into kinetic energy. As the ball rolls off the inclined plane and hits the ground, it comes to a complete stop, but only after it has passed all of its kinetic energy to the ground. And even though all of this is going on inside the system, the "losses" in energy – the energy I lost when pushing the ball, the potential energy the ball lost by rolling down to the ground, and the kinetic energy the ball lost to the ground coming to a complete stop – all exactly equal the "gains" in energy: the energy the ball received from my initial push, the kinetic energy the ball gained by being in motion, and the kinetic energy the ground gained from the ball as it rolled off the inclined plane and eventually came to a complete stop.

The energy is merely being passed from one object to another, while the total amount never changes. However, if the constraints are loosened a bit, things will change. Suppose heat is allowed to pass in and out through the walls. Further, we allow for work to be done on the system or by the system in a manner as previously described. Now, the system is interacting with its surroundings via *heating* and *working*, and the energy of the system will change.[61] Once again, since the formulism is independent of the exact details of the system, we have a very powerful tool that applies to a wide variety of systems.

For example, consider a glass of water with a lid covering the opening (thus preventing water molecules from escaping through evaporation). The glass and the cover form the boundaries of the system, and inside are all the water molecules. If the glass of water has come to *equilibrium*, it's now at room temperature, and heat is not passing in and out of the glass

---

[61] It's important to understand that we don't need to know the exact nature by which heating or working occurs in order to determine the energy change of a system. We simply need to determine the amount of heat or work leaving or entering the system. As to the details of what goes on within the boundaries of the system, this is of no consequence as long as we can make this determination.

– remember that you need a temperature difference for heat to flow (from hot to cold). Moreover, if it's just sitting there, no work is occurring, either.[62] In other words, our glass of water approximates an isolated system, and we expect, just as for our other system (inclined plane, rolling ball, and me) that the energy is not changing.

Now, unlike before, we have no way to make a detailed analysis of all that is happening inside. After all, we can't even see all the individual water molecules. To be sure, the water molecules are exchanging energy as they bump into each other, all the while conserving energy as the "losses" and "gains" perfectly balance each other.

Like Thomson, Clausius had also recognized the "energy concept." However, when he published his work in 1850, just slightly ahead of Thomson, he was less complete in his physical description and simply didn't understand the main ideas as completely as Thomson. So even though Clausius had beaten Thomson to the punch by almost a year, Thomson's description of the energy of a system and how it changes by its interactions with the surroundings was much more complete. Thomson originally called the energy of a system *mechanical energy*, but later, in 1856 he chose (the better name) *intrinsic energy*. Helmholtz would later call it *internal energy*.[63]

---

[62] If you were to stir the water, for example, you would be doing work on it. The process of stirring will change the energy by adding kinetic energy to the water.

[63] Specifically, internal energy (as used today) refers to the kinetic energy and potential energy of the molecules constituting our system of interest. It's the average value of both of these quantities that gives rise to the value we call the internal energy of a system. To understand more clearly why we take the average value, let's once again consider a glass of water as our system. The glass of water is made up of many individual water molecules that are moving around quite a bit at room temperature. This isn't something we can see with our eyes, but it's something that can be measured with experiments. As the molecules in the glass of water move around, their potential energy and kinetic energy will change, but the total energy won't if the system is closed as described earlier. But since the potential and kinetic energies of the molecules are constantly changing over time due to their motion, it's the average value (over time) of each of these quantities that is the most meaningful, with the sum of both these averages then defining the internal energy.

The Universal Tendency of Energy

In 1847, when Thomson learned of Joule's experiments (on the mechanical equivalent of heat) demonstrating that work could be converted to heat, he immediately recognized the impact of this discovery. It was clear, although not explicitly demonstrated by Joule's experiments (but nonetheless claimed by Joule), that this *equivalence* meant that one would expect the conversion of heat into work to be possible as well. This caused problems for Thomson since, at that time, he was still a proponent of caloric theory, which stood in direct opposition to Joule's conclusion. Nonetheless, Thomson resolved this struggle to full satisfaction when he published *On the Dynamical Theory of Heat*. But Thomson still struggled with another issue: What happens to the work Joule described when heat simply flows, or is conducted through an object, as it moves from a region that is hot to one that is cold?

By now, we thoroughly understand that one can use a heat engine to generate work from heat. But if instead we take this same quantity of heat and allow it to simply flow – bypass the heat engine altogether – it appears that the amount of work that would have been otherwise obtained has actually been lost; this troubled Thomson. To be sure, in the case of Carnot's reversible heat engine one gets the maximum amount of work possible from a given amount of heat, while, on the other hand, with Fourier's free conduction theory of heat, we are left with the minimum amount of work possible – in fact, no work at all would be produced. In the latter case, it's as if the available work has simply disappeared, and Thomson wanted to know exactly where it had vanished. He writes:

> The difficulty which weighed principally with me in not accepting the theory so ably supported by Mr Joule was that the mechanical effect stated in Carnot's theory to be *absolutely lost* by conduction, is not accounted for in the dynamical theory [of Joule] otherwise than by asserting that *it is not lost* ....

In 1852, Thomson published the short article *On a Universal Tendency in Nature to the Dissipation of Mechanical Energy*. In this article, Thomson states that, not all energy is created equally.[64] Some energy can be used to do work, and unfortunately, some energy can't. Consider the energy available in a river compared to that in the ocean. Needless to say, the ocean has much more energy than a river. You need only look out upon the ocean and watch the crashing of the waves upon the shore to be convinced. But how do you actually extract this energy and do work with it?

There are numerous challenges to this, and all of them have to do with the random (disordered) nature of the motion of the waves. Fluctuations in size, strength, direction, and duration of the ocean's waves all make it challenging to squeeze out the enormous amount of energy that is indeed available to produce work.[65] On the other hand, the mostly constant, steady, uniform (ordered) flow of a river makes it a much better candidate for energy extraction. This is why we build hydroelectric plants on rivers and not oceans.[66]

Thomson concludes that nature favors this random or dissipated type of energy; as such, once it has dissipated (as in the case of the ocean) it becomes essentially impossible to extract for useful work. In fact, if it's possible, and you do want to get this energy out, it will actually take work to do it. Indeed, this requisite work signifies the dissipation of energy as the direction favored by nature.

Thomson's Law of Dissipation alluded to a behavior of energy not accounted for in the first law. Consider Carnot's heat engine, where we are able to use only some of the inputted heat to create work and the rest is simply inevitably tossed to the surroundings. So even in the most perfect of scenarios, where the most efficient of all possible heat engines is used,

---

[64] In the introduction of the paper, Thomson divides energy into two types, which he calls "statical" and "dynamical." For Thomson, statical energy represented a higher quality of energy from which work could be produced, whereas dynamical energy was of a lower quality, which does not lend itself to producing work.

[65] To be sure, we do get some work out of the ocean, such as when one is surfing the waves.

[66] Another example is the energy available in the molecules that make up the air. There's a lot of energy available here as well, but once again the random motion of these molecules makes it impossible to actually extract this energy for work.

the universe still demands the waste or dissipation of some heat. There's simply no way around it since, as Thomson puts it, this is the "universal tendency." And if we make no attempt whatsoever to harness the flow of heat for work with any sort of heat engine, then all of it will be dissipated as described by Fourier's theory.

So, in either case an amount of heat will be dissipated but not lost. This dissipated heat goes into the random motions making up the matter it flows into – like the random motion of the ocean's waves. So, not all energy is equal, nature tends towards wasting (dissipating) energy as heat, and this wasted energy is neither lost nor destroyed, but is simply passed to the atoms constituting matter, thus becoming unavailable for work.

From these concepts comes the idea that "ordered" energy represents a higher quality than "disordered" energy since it can be used for work. Once again, consider our example of the ocean, with its disordered energy manifested in the chaotic motion of the waves, versus the higher-quality, ordered energy of the uniformly flowing river that can provide us with work. The comparison shows that dissipation of energy is a process of degradation of energy to that of a lower quality from its previous higher-quality state; a degradation from order to disorder.

The first law tells us that energy is neither created nor destroyed, but merely transformed from one form to another, and therefore it's conserved. However, Thomson's Law of Dissipation made it clear that so much more is going on with energy than the first law describes. Not only is energy conserved, but it has a tendency to dissipate. Moreover, this dissipation results in its degradation from higher quality (ordered) to lower quality (disordered). Thus, energy has a "preferred direction" to its dissipation, and reversing this direction requires an amount of work.[67] Indeed, Thomson's Law of Dissipation may have been his most important contribution to the area of thermodynamics. In fact, Thomson's Law of Dissipation provides a conceptual statement of *the second law of thermodynamics*.

---

[67] Merely supplying the required energy in the form of work won't reverse all irreversible processes; they are truly irreversible in this case.

CHAPTER 7

The Preferred Direction

Entropy as Nature's Signpost

Nature seems to have a "preferred direction" for certain processes. The heat from a cup of hot coffee will flow out to the surroundings, causing it to cool off. The creamer added to this same cup will mix in, regardless of whether it's stirred or not. Once overall equilibrium has been established, the coffee and the surroundings will be at the same temperature and the creamer will be uniformly mixed throughout.

As we all know, experience has taught us that heat won't suddenly flow from the surroundings back into the coffee, causing it to warm up again. Neither will the creamer suddenly unmix itself. And if we should knock the cup off the kitchen counter, it will most likely break upon hitting the floor. We can expect that (much to our disappointment), no matter how long we wait, the glass won't suddenly reassemble itself as it jumps back onto the counter. We'll have to clean this mess up ourselves. These processes, and many more alike, are called *irreversible* – they have a preferred direction according to nature's design, and the reverse direction is simply not the preferred one.

In effect, we're seeing that the flow of energy is governed by something in particular – something not described by the first law. In the midst of a changing vision about heat and from the desire of engineers to improve the efficiency of the heat engine, came the identification of a new quantity that serves as a counterpart to energy. It clarifies not only how much work one can get from a heat engine, but also why nature has a preferred direction (e.g., why heat flows from hot to cold). Moreover, this quantity is intimately connected to the world of atoms, and as such it's instrumental in our understanding of this *microscopic* world. This quantity is *entropy*.

Favored and Disfavored

By 1852, Thomson had come to believe that heat could be both transformed into work, as described by Joule's theory, and free-flowing to produce no work at all, as described by Fourier's theory. In the latter, heat was simply dissipated, but not lost, in accordance with the first law. Moreover, he distinguished between high-quality and low-quality energy and insisted that the universal tendency for energy is to dissipate as heat, making it unavailable for work. But Thomson wasn't the only one thinking about the finer points of heat.

In 1850, Clausius stated that the natural tendency of heat is to flow from high temperature to low temperature.[68] This intuitive and easily verified statement was Clausius' initial contribution to what would eventually become *the second law of thermodynamics* (or simply *the second law*). However, in 1854, Clausius set aside this simple statement and sought a rigorous mathematical formulation. The end result was the mathematical formulation of the second law and a new physical property,[69] which he would finally come to call *entropy* in 1865.

In 1854, Clausius reviewed everybody's favorite model system at the time: Carnot's reversible heat engine. Clausius noted that in a heat engine, heat undergoes two mechanisms available to it at the same time: *conversion* and *conduction*. Conversion is the process whereby heat is transformed

---

[68] Another way to say this is that the natural flow of heat is such that it cools an object down rather than heats it up. Heat flows from your coffee to the surroundings, resulting in its cooling. It does not naturally flow the other way (from the surroundings to your coffee, resulting in the spontaneous heating of your cold cup of coffee).

[69] Recall from Part I, a physical property of a system describes a unique physical feature of the system.

into work or vice versa, while conduction is the process whereby heat flows from high temperature to low temperature, or in the opposite direction.[70]

For each mechanism, Clausius asserted that there's a favored and a disfavored direction. Clausius considered the favored direction for conversion to be when heat is produced from work, as in Joule's experiments where heat results from the work of the falling weight and friction. Therefore, the disfavored direction, according to Clausius, is when heat is consumed to produce work, like in a heat engine. Clausius deemed the favored direction for conduction to be (obviously) the flow of heat from high to low temperature, and thus considered the flow of heat in the opposite direction to be disfavored. Whereas the favored directions simply occur as a part of nature's master plan, the disfavored directions need to be coerced.

For example, in order to get work from heat you need a device, like a heat engine; otherwise, it will simply dissipate without the production of work. However, the favored direction – or the one preferred by the universe – where heat is produced from work, occurs easily. It happens from any sort of effort as a result of friction. Whenever your feet slide across the floor as you walk, or the tires of your car rub against the ground as you drive, or your fingers bang on the keyboard as you type, among countless other scenarios, friction is right there producing heat from your efforts.

It's painfully clear that the favored direction of heat flow is from high to low temperature; it's why things cool down naturally but never heat up on their own. Moving heat from low to high temperature requires work to be done by a heat pump (as discussed earlier), such as an air conditioner or a refrigerator. Using these concepts and revisiting Carnot's heat engine, Clausius was able to come to an amazing conclusion.

Clausius realized that in a heat engine both conversion and conduction occur simultaneously. However, conduction occurs in the favored direction (the flow of heat from hot to cold), while the conversion occurs in the disfavored direction (the conversion of heat to work). Clausius reasoned that

---

[70] In 1851, Thomson was indeed also aware of these routes available to heat. He became focused on the process of conduction, which led to his work in 1852 describing the dissipative nature of heat.

in the idealized case of Carnot's reversible heat engine, these two processes are *equivalent*:

> ... these two transformations may be regarded as phenomena of the same nature, and we may call two transformations which can thus mutually replace one another equivalent.[71]

After establishing the mathematical guidelines for his new theory, Clausius calculates what he called the *equivalence values*: the equivalence value for conduction in the favored direction, and the equivalence value for conversion in the disfavored direction.[72] With a combination of impressive physical intuition and mathematical prowess, he found that if Carnot's heat engine is run in a cycle, the total of his newly defined equivalence values was – zero! Indeed, this may have been one of the most interesting zeros in the history of physics. Let me explain.

Running Carnot's engine[73] in a cycle merely corresponds to the physical way a real heat engine – like your car engine – works. At the end of every cycle, the engine returns to its original state or starting point, ready to begin the whole process again. However, it was the resulting zero in Clausius' theory that was the big deal. Specifically, it meant that Clausius had found a new physical quantity

---

[71] Specifically, what Clausius said is that either process in Carnot's engine can replace the other, once we have "reversed" it. For example, the process of heat flowing from hot to cold (its favored direction) can be replaced by the process of heat being converted to work (its disfavored direction) – once we have reversed it to work being converted to heat (its favored direction). This may seem confusing, but the point is that Clausius was trying to establish an overall relationship between the "favored" and "disfavored" directions. In doing so, he was able to develop a mathematical theory and moreover argue that these two seemingly different processes have a common physical origin.

[72] We now understand the mechanical (work) equivalent of heat is the amount of heat that one gets from doing a given amount of work, and indeed this is easily calculated by simply using an equation to convert the amount of work into the corresponding amount of heat. Clausius' equivalence value for the amount of work that one gets from a given amount of heat can be thought of as the "heat equivalent of work."

[73] Remember, Carnot's heat engine is a mathematical model, and what I mean by "run" in this case is merely proceeding with the calculation.

related to the mechanisms of heat (conversion and conduction). Not just any old physical quantity, but a very special type known as a *state function*.[74]

A state function is special because its value depends only on the present state of the system, rather than how the system acquired the state. A very familiar state function is volume.[75] For example, imagine filling a glass with water to exactly a volume of half full. Now, imagine filling it completely full but then dumping out just enough water to bring its volume to, once again, half full. Clearly, even though we achieved the volume of half full by two different routes, the end result is the same.

Moreover, the value of any state function will be unchanged if the starting and ending states of the system are the same – like they are in a cycle – regardless of the way we get from start to finish. Once again, consider the glass of water. Even though we obtained the volume of half full in two different ways, there wasn't a difference between the ending volumes; they were exactly the same: *zero*.[76] And that's what Clausius' zero meant too: some unknown state function related to heat remained unchanged as a result of Carnot's engine starting and ending in the same state. But what exactly was this state function? And what did it mean physically?

The Counterpart to Energy: Entropy

In 1854, Clausius identified a new state function by running Carnot's heat engine in a cycle. Surprisingly enough at this time, he didn't give it a name, but nonetheless he did define it as the "heat over temperature" for a given process. This could be, for example, the process of withdrawing heat from a heat

---

[74] The identification of a state quantity (or state function) as having a total value of zero when evaluated in a process "run" in a cycle is an excellent example of the use of mathematics to clearly identify meaningful physical quantities.

[75] Some other state functions include density, pressure, and temperature.

[76] Indeed, how strange would it be if we were only able to obtain the volume of half full by filling the glass to exactly that point each time, whereas if we were to overshoot it, simply dumping some of the water out would never lead us back to our desired volume of half full. Nonetheless, not all functions of the system are state functions, which means they do depend on the way the system obtains its state.

reservoir divided by the temperature of the heat reservoir, or the heat tossed to a cold reservoir divided by the temperature of the cold reservoir. That is:

$$\frac{q}{T}$$

where q is the heat, and T is the temperature in units of *Kelvin*.[77]

Recall that in Carnot's (reversible and perfectly idealized) heat engine the only heat lost to the surroundings is that which is tossed to the cold reservoir. This amount of heat is a required loss – there's simply no way around it – that the universe demands (nature's compensation) for the luxury of using a heat engine to do work. Nonetheless, Carnot's heat engine is still the most efficient heat engine there is, and of course we know it only exists on paper; you can't build Carnot's perfectly reversible heat engine.

A real heat engine is irreversible, which means it has other mechanisms by which heat is lost to the surroundings (and therefore not used for work), such as mechanical friction, heat lost through conduction, and other heat-generating mechanisms. As a result of all these other mechanisms, Clausius' sum of equivalence values, which was zero for a reversible process, ends up *always* being greater than zero for an irreversible process. It turns out that Clausius' sum is nothing more than the total over each q/T that makes up the process:

$$\frac{q_1}{T_1} + \frac{q_2}{T_2} + \frac{q_3}{T_3} + \frac{q_4}{T_4} + \ldots$$

So, this means for a reversible process run in a cycle, a given q/T has another q/T that cancels it out, sort of a partner if you will, thus ensuring that the total of all of them will be zero. However, for an irreversible process, this is not the case at all; each q/T does not have a partner, and the end result is a *leftover q/T* (or *uncompensated transformation*, as Clausius called it), which causes the total to be greater than zero; a positive value.[78]

---

[77] William Thomson (later Lord Kelvin) introduced the absolute temperature scale, where one adds 273.15 to the value of temperature as measured in degrees Celsius. On this scale, zero is defined as the absolute lowest temperature achievable.

[78] This is important so I hope you're reading this footnote. You only get a positive value when the convention is to treat heat absorbed or withdrawn (like from the

Processes in nature are irreversible and therefore have a *leftover q/T* here and there, which contributes (by adding that positive value to the total) to an ever-increasing amount of entropy in the universe.

A special type of irreversible process is a *spontaneous* process. A spontaneous process happens without needing any extra help to occur; it proceeds in the favored direction without the input of work. So, processes such as your coffee cooling down (as heat flows from it to the surroundings), a glass breaking upon falling to the ground (once the work of – for example – pushing it to the edge of the counter has been done), ice cubes melting in your drink, and many, many more such spontaneous process, all have this positive value. The larger this positive value, the more favored the process is, and the more it increases the entropy of the universe.

Clausius formulated many of the original ideas of his theory in 1854, identifying the remarkable new quantity q/T but choosing to keep it nameless. However, in 1865, after further deliberation, Clausius writes:

> I hold it to be better to borrow terms for important magnitudes from the ancient languages, so that they may be adopted unchanged in all modern languages, I propose to call the magnitude S the *entropy* of the body, from the Greek word [for] *transformation*. I have intentionally formed the word *entropy* so as to be as similar as possible to the word *energy*; for the two magnitudes to be denoted by these words are so nearly allied in their physical meanings, that a certain similarity in designation appears to be desirable.

Finally, Clausius concludes his paper of 1865 with his statements of the first and second laws respectively as:

---

hot reservoir or source) as a negative value and the heat that is given off as a positive. You will find that there's no agreement in the literature with respect to this convention. In fact, Clausius himself changed his convention between memoirs, introducing the one used here in 1854 and using it to write his famous *Clausius' Inequality* in 1862, only to switch to the opposite sign convention when writing in 1865 – at least he notes this shift in the latter memoir.

1. The energy of the universe is constant.
2. The entropy of the universe tends to a maximum.

So, whereas the universe keeps energy at a constant (energy is conserved), it continues to increase the entropy. Therefore, *no process that occurs will ever result in an overall decrease in the entropy of the universe.* The universe's tendency of maximizing entropy is reminiscent of "a universal tendency to the dissipation of mechanical energy" as stated by Thomson, and Clausius noted the connection.

Indeed, it's this dissipation – a loss of energy that might have been otherwise used for work – that separates a reversible process from an irreversible one. Evidently this is all part of nature's master plan to maximize the entropy of the universe, ultimately leading to its *heat death.* When the entropy of the universe finally reaches its maximum value, where it simply can't increase the amount of entropy anymore, all that will remain is energy dissipated as heat. This energy will be spread among all the atoms and molecules in the universe. This distribution of energy will be similar to before when we talked about the energy of the ocean (versus that of a river), in that we'll have no way to harness it for work. In other words, all the useful energy in the universe will be gone, and all that will remain is useless heat![79]

Clausius also related entropy to something he called *disintegration.* For Clausius, disintegration was a measure of the amount of separation between the molecules making up an object. Consider an ice cube that eventually acquires enough heat and melts to become liquid water. According to Clausius, as a liquid, the water has more disintegration than when it was ice. On average (over time), the molecules of water as a liquid are more separated from each other than as a solid, and this becomes more so when they are a gas.

---

[79] Moreover, it will be really cold. Temperature is a measure of the average *translational kinetic energy* that a single molecule in a given system possesses. The first law tells us the amount of energy in the universe is constant and therefore if I spread this fixed amount over all the atoms and molecules in the universe (my system in this case), a single atom or molecule won't have that much kinetic energy since there are so many atoms and molecules that make up our universe, and therefore the average will be quite low (almost zero), as will the temperature.

This was Clausius' attempt to provide a molecular interpretation of entropy. Nonetheless, Clausius stressed that the first and second laws were axioms and consequently were freed from the necessity of a molecular interpretation; a molecular interpretation was only supplementary.[80]

Clausius was a major contributor to the field of thermodynamics. His development of the idea of entropy and the second law that follows from it are profound concepts still taught and used in science today. (It's unfortunate his contributions often go unnoticed.) And while he was always thorough in the development of his theories, it wasn't enough to protect him from misunderstandings and harsh criticisms. Perhaps Clausius inadvertently brought some of this on himself.

Clausius' papers are often written in a verbose style, lacking a concise formulation of the pertinent ideas.[81] Things were complicated even further since Clausius depended on a considerable amount of intuition in drawing his conclusions. Moreover, the mathematics he used was rather unfamiliar, challenging even accomplished mathematicians such as Thomson and Maxwell. Perhaps worst of all, in his mathematical development of entropy he clumsily begins with one convention only to switch to the exact opposite in the end.[82]

Clausius' most persistent and outspoken critic was Peter Guthrie Tait (1831–1901). Careless in his reading of Clausius, Tait interpreted entropy as a measure of the energy available to do work. This is rather surprising given

---

[80] In the 1860s, molecular science was in its early stages, and Clausius didn't want to get caught up in speculative notions that might jeopardize his theory of entropy. So Clausius explored this line of thought and stopped, never arriving at a complete molecular interpretation of entropy.

[81] Our understanding of Clausius' work really comes from the clarification put forth by scientists such as James Clerk Maxwell and others.

[82] Specifically, Clausius was inconsistent in his convention for the heat absorbed (as from the hot reservoir) and the heat given off (as to the cold reservoir) in a process. In 1854, Clausius was quite content to deem the heat absorbed as negative and the heat given off as positive, but he switched to the opposite sign convention in 1865 when writing his major conclusions on entropy and the second law. This confusion still persists today in textbooks and notes on the subject of entropy. Thus, you need to be clear on the author's convention.

that entropy is not a form of energy in the first place, and even a passing glance at its mathematical form makes clear this fact, revealing the units of entropy to be those given by heat (which is energy, as you know) divided by temperature (in units of Kelvin).

And if it wasn't enough that in 1868 this blatant error appeared in Tait's book *Sketches of Thermodynamics,* it showed up again in a textbook published by (of all people) James Clerk Maxwell (1831–1879) titled *Theory of Heat,* which, aside from the error, ignored much of Clausius' work. After Clausius wrote a letter to the journal *Philosophical Magazine,* Maxwell noted his error and reconciled it in the second edition of his book, also (correctly) showing that entropy is actually connected to the energy that is *not* available to a system to do work. Today, we understand that it's the *free energy*[83] of the system that is the maximum amount of energy in the system that is available for work.

Entropy – as defined by Clausius – provides a counterpart to energy. While the energy of the universe is conserved, as insisted upon by the first law, its entropy tends towards a maximum. Energy and the first law that governs it can't explain why certain processes tend in what apparently is a favored direction; for that we need entropy. Nonetheless, its *thermodynamic* definition – "heat over temperature" – leaves something to be desired. We understand heat is a form of energy resulting from the motion of the tiny constituents that form matter (atoms). However, there's nothing in the thermodynamic definition that connects entropy – in any way – to atoms; there's no *microscopic interpretation.*

It took the amazing efforts of a few scientists to show that entropy does indeed have a microscopic interpretation, which not only complements the thermodynamic description originated by Clausius, but also extends it dramatically, making the entropy concept an even more powerful tool in modern-day science.

---

[83] In general, the *free energy* is simply a useful way to write the entropy of the universe in terms of the entropy and the internal energy (kinetic and potential) of just the system. The exact form of the free energy will vary depending on what macroscopic quantities we are holding constant, such as the temperature or pressure.

CHAPTER 8

The Other Side of Entropy

Entropy's Connection to Matter and Atoms

The move away from Clausius' thermodynamic description of entropy ("heat over temperature") and towards one involving the fundamental constituents of matter – atoms – has its beginnings in the study of gases and the desire to explain the behavior of these systems in terms of mathematical models. This area of theoretical study is known as *kinetic theory*, and its successor is called *statistical mechanics*.

Clausius was one of the initiators of kinetic theory, attempting to use it to put entropy on a molecular foundation, although insisting this approach to be only supplementary since, as far as he was concerned, the concept of entropy was an axiom firmly rooted in the second law. It was mostly due to the efforts of James Clerk Maxwell and Ludwig Boltzmann that kinetic theory and statistical mechanics really began to flourish, resulting in, among other things, the clear connection of the microscopic world of atoms to that of entropy.

Beyond Clausius' Entropy

James Clerk Maxwell was born in Edinburgh, Scotland, in 1831. His family moved to a small country estate in Middlebie, Galloway (southwestern

Scotland), that his father, John Clerk, inherited (the addition of the surname "Maxwell" was required to satisfy legalities of this inheritance). When he was eight, James' mother died of abdominal cancer; she was forty-eight. John Clerk Maxwell was an attentive and perhaps overly protective father. Unfortunately, he made the mistake of entrusting James' early education to a tutor who employed beatings as a teaching tactic. Fortunately, a visit from his maternal aunt, Jane Cay, discontinued this abusive treatment, as she was able to convince John Clerk to allow the younger Maxwell to continue his education at Edinburgh Academy. His initial experiences at the academy weren't so pleasant, either, as he endured bullying (his strange dress and strong Galloway accent made him an easy target) and acquired the nickname "Dafty," similar to "weirdo" as used today. Nonetheless, Maxwell endured and even made lasting friendships.

As an enthusiast of science and mathematics himself, Maxwell's father was very encouraging of his son's passion for science, and together they often attended meetings of the Edinburgh Society of Arts and the Edinburgh Royal Society. At the age of fourteen, Maxwell wrote a paper on a new method for constructing ovals. Although Descartes had already described a majority of this work, some of Maxwell's paper was original. Maxwell's father brought the work to the attention of James Forbes, a professor of natural philosophy at the University of Edinburgh, and this propelled Maxwell's career.

In 1847, at the age of sixteen, Maxwell began attending the University of Edinburgh, where he studied with Forbes and William Hamilton. As sworn enemies, Forbes and Hamilton pretty much disagreed on everything. However, agreeing Maxwell was gifted, they gave him the special attention they felt he deserved.

A skilled experimentalist, Forbes gave Maxwell free rein of his laboratory, while Hamilton, a philosopher, honed Maxwell's conceptualization skills by stressing the use of idealized models to provide insight into real phenomena. Although after his first term Maxwell had the opportunity to attend Cambridge (a favored choice at the time for those pursuing mathematics), he stayed and completed his undergraduate education at Edinburgh. Maxwell didn't find the classes very challenging and found time to pursue many topics on his own, such as his experiments with polarized light.

In 1850, he attended Peterhouse (a constituent college of the University of Cambridge in England) but after his first term transferred to Trinity College (another constituent college of the University of Cambridge). It was during this time that Maxwell began to fit in (despite his quirky sense of humor and eccentricities) and developed a number of friendships. In 1854, Maxwell graduated from Trinity with a degree in mathematics. He remained on the staff at Trinity for two more years, receiving a fellowship during his last year.

In 1856, a professorship opportunity became available at Marischal College in Aberdeen, Scotland. Among other things, this would provide a good opportunity to be near his father, whose health was now declining. His father assisted him in preparing the needed references for the position but died just prior to Maxwell's learning he had the job.

Maxwell was now twenty-five and a decade and a half younger than any other professor at Marischal. Although his written works (papers, formal lectures, and books) exemplified clarity, Maxwell was unsuccessful as a teacher, unable to convey lecture material to his students. In 1860, reorganization left Maxwell without a job, so he accepted an appointment at King's College in London. His five years here were perhaps the most creative of his life, and his work gave him much satisfaction. Of Maxwell's many contributions, what interests us here is his work on the motion of atoms.

Today, we take it for granted that all matter consists of tiny, invisible building blocks called atoms. Even before their existence was confirmed experimentally, scientists and mathematicians were using atoms in their physical models, imagining them in their mind's eye as they attempted to clarify everyday phenomena. Gases were among the first of the states of matter to which atomic models were applied.[84]

An early approach to the study of gases was known as *kinetic theory*, the name reflecting the fact that the motion (hence kinetic energy) of the

---

[84] Hero of Alexandria (c. 10–70 AD) maintained that the compressibility of air was a consequence of its being made up of tiny particles.

particles was the very thing under consideration. Specifically, the goal was to understand how the motion of the atoms in a gas results in the everyday properties we observe. Indeed, a container of gas – say, a balloon filled with helium atoms – is a challenging system to develop a useful model for.[85] The reality is that at room temperature and pressure[86] you have some 10^{23} (100, 000,000,000,000,000,000,000) atoms in that balloon,[87] and on average they are whizzing around at approximately 2800 miles per hour (even faster as the temperature increases)[88] with a given atom undergoing approximately a billion collisions every second with other atoms. Nonetheless, all this motion, as chaotic as it is, has a definite relationship to the overall properties of the gas.

That the properties of gases could be explained by particles in motion had been advocated in 1738 by Daniel Bernoulli (1700–1782), who proposed a very similar model to the one in acceptance today. However, at the time, Bernoulli's model had some impossible challenges to overcome. First, the idea that matter was made up of tiny, invisible particles was still not mainstream. Moreover, Bernoulli's *particles in motion* were at odds with Newton's *particles at fixed locations*, as described in *Principia* (1687), and Newton was hard to beat. Bernoulli also advocated the idea that the motion of particles produced heat, which majorly conflicted with the then widely

---

[85] Useful models correctly describe known experimental information and, as a bonus, correctly predict new behavior; they are descriptive and predictive.

[86] I am using 21°C (about 70°F) for temperature and 1 atmosphere for pressure.

[87] I am considering the popular balloon size of ten inches for the diameter.

[88] One example of atoms moving faster as the temperature increases comes from your car tire. As you drive your car, the friction between its tires and the road produces heat, causing the temperature of the air inside the tire to increase. This increase causes the molecules of air (air is a mixture of molecules – mostly nitrogen and then oxygen, together making up about 99% of air – and molecules are made up of atoms) to move faster, which means they collide more often with each other and with the wall of the tire (the container that holds them). This increase in the number of collisions in time results in increased tire pressure. This is easy enough to measure: simply measure the air pressure before and after you drive. This is also why a slow leak is not readily noticed right after you drive but shows up the next time you drive your car, such as driving to work the next morning after it had sat all night.

accepted caloric theory. So with three strikes against him, and being about a century too early, Bernoulli's theory failed to catch on, despite the fact that it correctly explained a well-known experimental relationship between the pressure of a gas and the volume of the container holding it (known as *Boyle's Law*). Before appreciating Maxwell's contributions, we need to discuss the very notable ones made by Clausius.

In 1856, August Karl Krönig (1822–1879) published a short memoir on kinetic theory that inspired Clausius to begin his own efforts on the topic. Clausius went well beyond Krönig's work and published a paper in 1857. A key result in Clausius' paper is an expression relating the pressure of a gas to the volume of its container and the average speed for the atoms making up the gas.

The power of such a relationship is this: it relates a *microscopic* quantity (the average speed of the atoms of gas) to *macroscopic* quantities (pressure and volume). In other words, it relates the behavior of the invisible world of atoms to the visible world via things we can measure, in this case pressure and volume.

For example, it's easy to measure the air pressure and determine the volume (written on the side) of your car tire. Now, with a relationship very similar to the one derived by Clausius, the number of gas atoms in the tire is also easily determined, even though they can't actually be seen. This was pretty powerful stuff indeed.

Clausius also suggested that the speed of the atoms in a container of gas takes on a *distribution*. In other words, not all the gas atoms move at the same speed, but each atom takes on a speed in a specific range. Clausius never actually determined this range. Rather, he removed its necessity from his calculations by treating each atom as moving at an average speed. In doing so, Clausius brought a tremendous amount of manageability to a seemingly unapproachable problem. The determination of this range of speeds and its importance was left to Maxwell.

What led Maxwell to kinetic theory was, oddly enough, Saturn's rings. In 1855, the topic of the biennial Adams Prize for mathematics at Cambridge was "The Motion of Saturn's Rings." At the time, it was unclear whether the rings were solid, liquid, or gas. Indeed, the true nature of Saturn's rings had been a fascination since Galileo first looked at them with his telescope in 1610.

While in Aberdeen, Maxwell devoted much of his time to the problem and in a letter to Thomson describes the rings as "a great stratum of rubbish jostling and jumbling round Saturn without hope of rest or agreement in itself …."

Maxwell constructed a theory that showed Saturn's rings couldn't be solid, liquid, or gas, but rather were made of many small, solid, colliding particles orbiting the planet, which were dynamically stable and provided a solid-like appearance. The solution won him the prize. Today, we do know that Saturn's rings consist of tiny rocks that collide with each other as they orbit the planet. Maxwell was now ready to think about another system of colliding particles: a container filled with gas atoms.

In his essay on the rings of Saturn, he commented on the challenge of constructing a mathematical model that could appropriately deal with the collisions of so many particles. Maxwell later noted that Clausius' work on kinetic theory provided a starting point for such an endeavor. However, whereas Clausius chose to treat each atom in a gas as traveling at a fixed (average) speed, Maxwell was determined to find the actual distribution (or range) of speeds available to a collection of gas atoms.

In the nineteenth century, most scientists believed that such a distribution of speeds (if it even existed at all) was simply an indication that the system of gas atoms hadn't quite "settled down" yet. In other words, the system wasn't at equilibrium, but rather was in a *nonequilibrium* state. The assumption was that, given enough time, the system would eventually come to equilibrium, and the gas atoms would then all be traveling at the same speed.[89]

The Speed of Atoms

Maxwell's initial efforts in kinetic theory began while still a professor at Marischal College. Maxwell wrote to fellow physicist and mathematician,

---

[89] To be sure, Clausius didn't believe this, but he did maintain that for the sake of calculation – even though there was a distribution of speeds for the atoms of gas – one could simply use an average speed. This led Clausius to a result for the pressure of a gas that was not quite right, which was later pointed out by Maxwell upon determining the mathematical form of the distribution of speeds in question.

George Gabriel Stokes (1819–1903), on May 30, 1859, describing the work as an "exercise in mechanics." Apparently, Maxwell had serious doubts about a kinetic theory that correctly described the properties of gases. Nonetheless, in 1860, he published a revolutionary paper, proving what Clausius had only speculated about: the speeds of gas atoms come from a specific distribution when the system is at equilibrium.

The condition that the system is at equilibrium is an important one. In this case, it means that the temperature, the number of gas atoms, and the volume of the container holding them remain unchanged; they're all constant. Maxwell realized that, for all practical purposes, the collisions occurring between the atoms in a gas result in an overall *random* motion. Therefore, much to the frustration of his contemporaries, he completely ignores the explicit collisions occurring between the gas atoms (and the walls of the container) and instead appeals to *probability theory*. The laws of probability govern such things as the odds of winning the lottery being very small; the next turn of a card giving you twenty-one (or not) in a game of blackjack; a flipped coin giving heads (or tails) 50% of the time (on average), and the like. Maxwell showed that these same ideas apply to understanding the properties of a very large number of colliding gas atoms.

Maxwell considered the velocity of the atoms of a gas in equilibrium. For the velocity, a vector, it's necessary to consider both the amount and the direction of the motion. By assuming that each atom of a gas has the same likelihood, or probability, to move in any of the available directions,[90] he showed that the atoms of a gas do indeed have a range of velocities available to them. Therefore, they don't all travel at the same speed; rather, a very small number travel very slowly or very fast, while the majority travel at more intermediate speeds. In other words, according to the *Maxwell distribution*, when the system is at equilibrium there's a higher probability of finding an atom in a gas traveling at an intermediate speed than either at very slow or very fast speeds. Now, in practice we find that the Maxwell

---

[90] There are three unique directions available to a gas atom, called x, y, and z, when using Cartesian coordinates. They describe the familiar motions of "up or down," "left or right," and "forward or backward."

distribution applies to a gas that behaves as an *ideal gas*, where the atoms of the gas don't experience the attractive ("pull") and repulsive ("push") interactions that a real gas does.

In 1860, Maxwell needed merely a single page to derive this amazing result, which also would allow him to calculate other important properties of a gas that matched experimental observation. To be sure, a system that is not in equilibrium will also take on a distribution for the velocities of the gas atoms. However, there's a catch: it won't be a Maxwell distribution.

Consider our system of gas atoms once again, where the number of atoms and the volume of the container are unchanged, but the temperature is changing in time. Let's assume our nonequilibrium system is evolving towards equilibrium, and therefore the temperature will eventually be constant. As the system evolves, so does its velocity distribution, and until the system finally reaches equilibrium, the distribution will continue to change over time. However, once the system has reached equilibrium, the velocities will acquire a Maxwell distribution. This is regardless of the initial state of the system or how it was initially prepared. Finally, the tendency of our nonequilibrium system to evolve towards equilibrium is a spontaneous process, which means the entropy is increasing irreversibly along the way.[91] The evolution of our system to a Maxwell distribution at equilibrium plays a vital role in understanding certain physical processes.

Consider a hot bowl of soup. Blowing on a spoonful prior to placing it in your mouth does cool it down. This is because the soup is made up of atoms moving at a range of speeds. The atoms with the higher speeds (the hotter ones) are on top, hovering above the atoms with the lower speeds (the cooler ones). So, the end result of blowing on the spoonful of soup is that the hotter atoms are blown away, while the cooler ones are left behind. If the atoms in the soup were all traveling at the same speed, there would

---

[91] However, this doesn't mean it increases in the "absolute" sense, that every single step along the way results in an increase in entropy. Rather, it occurs only in the statistical sense. That is to say that some steps result in an increase, while others result in a decrease. The overall change in entropy when the system finally reaches the equilibrium state from the initial nonequilibrium state will be an overall increase. The second law guarantees this fact.

be no hot or cold atoms; they would all be the same. In this case, you could blow until you're blue in the face, but the temperature of the soup would never change.

Before Maxwell, probability and statistics had been used for data analysis (in the social sciences and physics). However, Maxwell's approach distinguished itself because it used these methods to correctly describe the actual physical process itself. Maxwell's work in thermodynamics was done over a period of years, and yet he never wrote any major papers on the topic. We know of his efforts from his correspondence with Thomson and Tait, and from his *Theory of Heat*, first appearing in 1870 and running for eleven editions thereafter. In addition to his own contributions, Maxwell was instrumental in clarifying the ideas of Josiah Willard Gibbs (1839–1903), Boltzmann, and Clausius, and in reconciling the disparate views among Clausius, Tait, and Thomson.

Aside from his eccentric character, Maxwell was generous, unselfish, and had a deep sense of duty. When his wife, Katherine, was sick (she was often in frail health), Maxwell sat by her bed for three weeks and attended to his responsibilities at his laboratory. All the while his own health was failing. When reviewing another's paper for publication, Maxwell's comments often provided more insight into the topic than the paper itself. His comments on a paper by William Crookes (1832–1919) offered advice that, had it been pursued, might have led to the discovery of the electron. Perhaps surprising was Maxwell's sense of the metaphysical. In a letter to a friend he writes, "that the relation of parts to wholes pervades the invisible no less than the visible world, and that beneath the individuality which accompanies our personal life there lies hidden a deeper community of being as well as of feeling and action."

On November 5, 1879, Maxwell died at the age of forty-eight in Cambridge of abdominal cancer (the same cancer that killed his mother at the same age). Maxwell's statistical approach to the dynamics of gases opened the door for Ludwig Boltzmann to arrive at a microscopic interpretation of entropy, thus going far beyond Clausius' thermodynamic definition of entropy of "heat over temperature."

Entropy and Probability

Ludwig Boltzmann (1844–1906) was born in Vienna and attended the University of Vienna, where he received his doctorate in 1867. Boltzmann was a restless spirit, changing (by choice) from one academic position to another for a total of seven times in his almost forty-year career.

From the early 1870s on, Boltzmann was a scientific superstar and very much in demand. To get Boltzmann to accept a professorship of theoretical physics at the University of Vienna in 1894, the Austrian minister of culture had to offer him the highest salary then paid to any Austrian university professor. Boltzmann had already been a professor at the University of Vienna twice before, once from 1867 to 1869 as an assistant professor, and a second time from 1873 to 1876 as a professor of mathematics. Nonetheless, in 1900, he left for a third time.

His final return to Vienna was in 1902, upon which he succeeded none other than himself as professor of theoretical physics; the position had been vacant since his departure two years before. Well aware of his wandering tendencies, the Austrian authorities allowed Boltzmann's return conditional upon his promise never to take another job outside of Austria, a promise he kept. Boltzmann's personality was such that he would fluctuate between extreme highs and lows, and he himself joked about this duality in behavior, likening it to the fact that he was born between Shrove Tuesday and Ash Wednesday – between celebration and penance, so to speak. Today, Boltzmann would most likely be diagnosed as bipolar (he had other health issues: asthma, migraines, poor eyesight, and angina).

Boltzmann was a theoretician through and through (making it a curious fact that he was a professor of experimental physics at the University of Graz from 1876 to 1890, which was his longest stay at any of his positions). He says of himself: "I am a theoretician from head to toe. The idea that fills my thoughts and deeds [is] the development of theory. To glorify it no sacrifice is too great for me: since theory is the content of my entire life."

To be sure, Boltzmann was probably only second to Maxwell as a theoretician. But Boltzmann was not only famous for his theoretical prowess but also his exceptional success as an educator, something never accomplished by Maxwell. Moreover, he was very attentive to his students and treated them as peers, allowing informal discussions where they were permitted to ask questions and even criticize him.

In 1866, at the age of twenty-two, Boltzmann wrote his first paper of any significance titled *On the Mechanical Meaning of the Second Law*. Here he contrasted the ambiguous nature of the second law against the very secure stature of the conservation of energy described by the first law. He attempted to provide a general proof of the second law and relate it to classical mechanics as described by Newton.

Although Maxwell didn't formulate a theory of the second law himself, he regarded this approach as fundamentally misguided since for him the second law was purely probabilistic or statistical in nature, and therefore had no basis in a purely classical mechanical description. Boltzmann was not alone in his misdirection. Clausius also attempted to bring the second law to terms with the principles of mechanics. Boltzmann soon realized this approach was fatally flawed, whereas Clausius, even after several attempts, was never convinced.

Clausius' first attempt was in 1862 with his introduction of "disintegration," which was supposed to be a measure of the amount of separation between the atoms (or molecules) in an object. In 1870 and 1871, he published papers that claimed to provide a mechanical view of the second law. His latter paper prompted a quick response by Boltzmann, who in his paper pointed out that Clausius had essentially reproduced the efforts of Boltzmann's 1866 publication. In a following paper in 1872, Clausius graciously conceded, stating the "extraordinary demands" on his time had made it difficult for him to keep current with the scientific literature. Nonetheless, all these treatments were flawed, and although Clausius never came to terms with the statistical nature of the second law, Boltzmann did and was able to demonstrate this by extending the work started by Maxwell.

In his paper of 1866, Boltzmann made no mention of Maxwell's work, which is more likely because he was still learning English (in particular so he could read Maxwell's original work on electromagnetic theory) than an oversight on his part. However, just two years after this failed attempt, and now very familiar with Maxwell's work on the distribution of the speeds of atoms in a gas, Boltzmann had what he was looking for: an explanation of entropy and the second law in terms of probability and statistics.

Recall that the Maxwell distribution describes the range, or distribution, of speeds for the atoms in an ideal gas in equilibrium at a given temperature. We have already learned that "energy of motion" is kinetic energy, and something that has a speed must be in motion and consequently have kinetic energy. Therefore, the Maxwell distribution is also describing the kinetic energy distribution of the ideal gas atoms at equilibrium. To put it another way, the kinetic energy distribution of the entire system of an ideal gas at equilibrium is described by the Maxwell distribution – this is what Boltzmann realized. In 1868, Boltzmann was able to show that the idea of a distribution could be extended to the *total energy* of a system at equilibrium, both the kinetic and the potential energy contributions. In fact, with his more general approach to the problem, Boltzmann was also able to obtain the Maxwell distribution for a system of ideal gas atoms at equilibrium.

So, while the gas atoms of a given system are zipping around, colliding with each other and with the walls of the container that hold them, when the system is at equilibrium, the system as a whole is sampling a range of total energies given by the *Boltzmann distribution*. Each of these energies describes the system as being in a certain *microstate*. Specifically, a single microstate is described by all the positions of the gas atoms and their respective velocities at a given instant in time. As time goes by, the system will move from one microstate to another and given enough time, the system will sample all the microstates available to it. However, it won't sample each microstate with the same frequency; it will sample the more probable microstates more often. According to the Boltzmann distribution, the microstates with a lower total energy are the more probable ones, with the probability of their occurrence given by the *Boltzmann probability*.[92]

---

[92]   In the case of a system at constant (same) energy, things become a bit simpler. The system will still have a set of microstates, but each microstate will be the same energy. The Boltzmann distribution won't be a range of energies but only a single energy. Therefore, the probability of a given microstate is just one over the total number of microstates constituting the system; in other words, every microstate has the same probability.

The Boltzmann distribution, the Boltzmann probability, and micro-states are fundamental ideas of *statistical mechanics*, which allow us to correctly calculate certain things about a system – and not just a system of gas atoms, but for all *classical* (as opposed to *quantum*) systems. Things such as pressure and temperature can be calculated, sometimes with the help of computer simulations. So, while we can't see atoms, their crazy motions, or the microstates that result,[93] with the methods developed by Boltzmann and others around these concepts, we can correctly describe many of the things that we observe in our daily lives.

When you look at something, what you're seeing is its physical state, or *macrostate*.[94] Again, consider a balloon filled with air. You can't see the air molecules moving around as they collide with each other and the walls of the balloon. What you can see is the shape (volume) of the balloon, and you can measure the temperature. In this case, the macrostate of your system[95] is well described by properties you can see and measure: temperature and volume.[96] However, the microstates, resulting from the colliding molecules, are hidden from view.

---

[93] However, these things can be seen using computer simulations. Here, we can view the motion of the atoms and their trajectories and all the microstates that result. Moving the atoms according to Newton's equations of motion and following the tenets of statistical mechanics, we can calculate a variety of properties of many systems and find good agreement with experimental results.

[94] I am taking the time to introduce and clarify the terms *microstate* and *macrostate* so there will be no confusion. Oftentimes, authors will simply say "state" and you will have to figure out which one they are talking about from the context, and sometimes they will use "state" incorrectly altogether.

[95] More correctly, the macrostate of a system is given by the *variables* of the system that don't change even though the atoms of the system are moving around – that is, even though the system is changing microstates. In my example, I said the temperature and volume (shape) of the balloon, which you can see, were the macrostate variables for the balloon. However, the number of atoms in the balloon would be the other macrostate variable. Clearly, even though you can't see the atoms, the number confined to the balloon is not changing – unless there's a leak.

[96] The number of air molecules inside the balloon is also a quantity associated with the macrostate. This may seem confusing since you can't see the molecules. Nonetheless, you do know that the total number is not changing inside the balloon unless of course you have a leak.

So, in the end, the many microstates are the *hidden* states of the system, while the single macrostate is the *overall* state with the physical properties we see and measure. In some sense, the macrostate is the coarse-grained, or fuzzy, version of the system.

It's kind of like this: when I was kid, I liked shaking the boxes containing my Christmas presents. As I shook a box, the contents inside would jiggle around (sometimes break) and, in essence, move from one "microstate" to another. Now, I could never see all this jiggling around (and all the different "microstates"), but what I could see in the midst of all the shaking – and what never changed – was the nicely wrapped box with the bow on top: the "macrostate" was always the same even though the "microstates" inside the box were changing with each shake.

The distinction between a macrostate and all the microstates constituting it leads to a more fundamental understanding of entropy, more than just as "heat over temperature." Boltzmann showed us that the more microstates a system has available to it, the higher its entropy.[97] Recall that a spontaneous process occurs without any outside help or work input; it just happens. Clausius taught us a spontaneous process occurs because it's *entropically favored*; this is the direction that leads to an increase in entropy and is therefore the favored direction.

Using Boltzmann's concept of microstates, we can also say this: a spontaneous process occurs because this is the direction that leads to more microstates. The quintessential example of such a process may be mixing. Consider what happens when cream is poured into coffee: it spontaneously mixes. We may have imagined the cream huddling itself at the top of the cup, never fully mixing throughout. However, the second law assures us that the natural tendency will be for the coffee and cream to increase their overall entropy.

Thus, rather than huddling at the top of the cup, the cream moves through the coffee. This process of *diffusion* gives the cream access to much

---

[97] This is so important that if you remember nothing about microstates and macrostate, at least remember this relationship of microstates to entropy: when a system has more microstates, it has higher entropy.

more space within the cup than it would have otherwise had by simply remaining at the top. Moreover, the space at the top of the cup, previously occupied exclusively by the cream, is now also available to the coffee. The process of mixing has afforded both the cream and the coffee access to more available space within the cup. In this case, more space for both means more microstates overall, resulting in the entropy of the system being maximized.

So, does this mean that you will never see the creamer spontaneously unmix? In short, yes. The reason for this is that there's essentially only one microstate where the cream and coffee are fully unmixed, compared to the many more microstates where they are (at least to some extent) mixed. In the end, it has to do with probability: having more ways for one thing to occur (mixing) than another thing to occur (unmixing) favors the one with the most chances. In this case, there are more chances to find the cream mixed with the coffee than there are to find it unmixed (just like there are so many more ways for you to lose the lottery than to win).

Now, Boltzmann never said there was *no* chance. To be sure, there's a nonzero chance that one day the coffee and cream will end up unmixed. But the chance is so small because, once again, there are so many more ways for them to be mixed. So, as the atoms in the coffee and creamer zip around, colliding with each other and the cup, they spend most of their time in *microstates*, which result in the overall *mixed macrostate* (physical state) we see.

Let's look at one last example, something perhaps a bit more tangible than atoms zipping around, which creates microstates that you can't even see in the first place. Let's consider a deck of cards.

When you buy a new deck of cards, the first thing you will notice is that all the cards are in order; each suit is arranged in sequence from ace to king. In terms of what we have been talking about, this would constitute an "ordered" microstate for the deck. Now, imagine we take this deck of cards and place it in an automatic shuffler. As the cards are shuffled, the deck moves from its originally ordered microstate to another microstate, then another, and so on. We can imagine something similar in a gas, where the constant movement of the atom causes the system to "shuffle" between microstates.

Now, we stop the shuffler and look at the cards. We find the current microstate is "disordered." That is to say the suits are all mixed up and out of sequence, nothing like before the deck was placed in the shuffler. We place the cards back in the shuffler and continue the process of shuffling, removing, and looking at the resulting microstate. Surely we would expect, from experience alone, that each time we observe another microstate it's disordered. In fact, after all this shuffling, it's likely we'll never find the system of cards back in the original ordered microstate they started in. It's not that it's impossible, but rather that there are so many more ways to find the deck in a disordered microstate, compared to only one way to find it in the ordered microstate. It's clear that the deck of cards has many more ways to end up in a disordered *microstate* than an ordered *microstate*. Therefore, the disordered *macrostate*, or *phase*, has higher entropy than the ordered macrostate, or phase, and is thus favored.

Nature behaves this way in general, not just a deck of cards. Consider water starting off as ice (solid), melting to become a liquid as it acquires heat from the surroundings, and then becoming steam (gas) upon further addition of heat. In each subsequent phase, the addition of heat results in increased motion of the molecules of water. As a solid, the molecules of water essentially don't move. However, as the ice begins to melt, the molecules of water begin to jiggle around more, and even more so when they are completely in the liquid phase. Finally, with the addition of even more heat, the water molecules go from their jiggling motion to outright collisions between each other as they enter the gas phase. The increase in motion, going from solid, liquid, and then to gas, results in more microstates for each subsequent phase. Therefore, as it was for the deck of cards, more microstates mean more entropy.

Often the increase in motion of the molecules is associated with an increase in the degree of disorder of the system. Therefore, the solid phase is the most ordered, and the gas phase is the most disordered. This perception is more of a convenience that also parallels well with our deck of cards analogy. However, it overlooks the further-reaching physical phenomenon of what's occurring: the addition of energy in the form of heat (q) has allowed the system to increase its microstates (and in doing so, change phases in the example we discussed). In fact, if we denote the change – or in this case the increase – in entropy as ΔS, we can write:

$$\Delta S = \frac{\Delta q}{T}$$

which is the thermodynamic expression we saw before.

In Boltzmann's time, the idea of matter being made up of small, invisible particles, which today we call atoms, was not yet mainstream. So, you can imagine that Boltzmann's theories, which presumed their existence, might have met with some resistance. Boltzmann had harsh critics but he refused to back down. In 1898, Boltzmann published the second volume of his *Lectures on Gas Theory*. Here we can get a sense of his perseverance for the existence of atoms:

> In my opinion it would be a great tragedy for science if the [kinetic] theory of gases were temporarily thrown into oblivion because of a momentary hostile attitude toward it …. I am conscious of being only an individual struggling weakly against the stream of time. But it still remains in my power to contribute in such a way that, when the [kinetic] theory of gases is again revived, not too much will have to be rediscovered.

Among Boltzmann's most prominent adversaries was Friedrich Wilhelm Ostwald (1853–1932). Ostwald was against all theories in chemistry and physics that involve the use of atoms, especially the kinetic theory of gases. For Ostwald the unifying concept in science was energy. For him entropy described the dissipation of energy (similar to Thomson's conceptualization) and had absolutely nothing to do with the motion of atoms resulting in microstates of a system, as described by Boltzmann. For Ostwald atoms were merely artifacts of mathematical theories, not manifestations of physical reality.

Ernst Mach (1838–1916) was another anti-atomist, although not a subscriber to a unifying energy concept as advocated by Ostwald. Mach simply couldn't accept the idea of atoms since he saw no supporting evidence for their existence. With regards to Boltzmann's microscopic interpretation of entropy, Mach wrote, "In my opinion, the roots of this [entropy] theorem lie much deeper, and if the molecular [atomic] hypothesis and entropy theorem be brought into harmony, it is a piece of good luck for the hypothesis but not for the entropy theorem."

During the summer of 1905, Boltzmann gave a series of lectures at the University of California, Berkeley. He was at the height of his fame. Students packed in to see his lectures, and colleagues sought his scientific advice. Upon his return to Vienna (unaware of the vindicating paper written by Einstein in 1905), Boltzmann recounted his journey in a piece called *A German Professor's Journey into Eldorado*.[98]

In early 1906, Boltzmann was on vacation with his wife and daughter near the Italian seaside town of Trieste. Once again struggling with depression, Boltzmann tied a short rope to a window crossbar and hung himself, while the women were off swimming. His daughter returned to find him dead. Ludwig Boltzmann is buried in the Central Cemetery of his native Vienna. On his tombstone is Boltzmann's equation for entropy as it relates to the number of microstates of a system:

$$S = k \ln W$$

where S is the total entropy of the system, k is Boltzmann's constant,[99] and W is the total number of microstates corresponding to a given macrostate of the system.

Beyond Heat Engines

When speaking of entropy we can consider two complementary versions: *thermodynamic* and *statistical mechanical*. The original description of

---

[98] At the time of Boltzmann's visit to Berkeley, the drinking and buying of beer and wine, much to Boltzmann's displeasure, was forbidden. Nonetheless, Boltzmann soon learned of a shop selling wine in Oakland, California, and apparently made frequent visits thereafter.

[99] Boltzmann's constant is used to convert temperature (in Kelvin) to units of energy. Boltzmann never introduced this constant but wrote this relation as a proportionality instead. Planck later introduced the constant.

entropy is purely thermodynamic and is due to the work of Clausius. He discovered that when running Carnot's reversible heat engine in a cycle a very interesting quantity emerged: "heat over temperature." Eleven years later, he came to call this remarkable quantity entropy. The most important thing Clausius noticed was that for irreversible processes, the entropy always increases.

Irreversible processes are interesting to us because these are the processes we experience every day. Irreversible processes are not reversible without some sort of work input (if at all). A glass that has fallen to the ground and breaks is an irreversible process; the entropy has increased upon breaking, and the glass won't suddenly reassemble without some sort of work on our part – perhaps gluing it back together – and even then it won't be back to its original form. Moreover, the breaking of the glass occurred *spontaneously*.[100] That is to say, aside from the bit of work it took to set it in motion (perhaps knocking it off the kitchen counter), the whole process occurred on its own. Thus, we can see that increasing entropy indicates a "favored direction" in nature. Upon seeing the broken glass, we would all agree that at some point in the past, it was unbroken, or in other words at some point in the past the entropy was lower and now it's higher. Therefore, entropy seems to have some relationship to the direction of time, or *time's arrow*.

The first law tells us that energy is always conserved, merely being transformed from one form to another, and the second law tells us that entropy is not conserved at all but tends towards a maximum. Indeed, it seems nature is constantly producing entropy. No matter how efficient any given machine is, it will still produce entropy. It's nature's law.

When you drive your car the engine will lose heat to the environment simply because nature demands it in the process (nature's compensation for letting you use a machine to do work). Therefore, it's impossible to use all of the fuel (energy) for work. Moreover, there will be additional heat

---

[100] Not all irreversible processes are spontaneous, but all spontaneous processes are irreversible. Most processes in nature are irreversible. The only reversible processes in nature are systems in equilibrium.

loss because of friction in many different forms in your car's engine. And since entropy, according to Clausius, is "heat over temperature," nature has once again found a way to ensure entropy is tending towards a maximum.

While Clausius' definition of entropy is quite correct, it's rather incomplete. Clausius himself tried to find a microscopic relationship for entropy, one that would tie it to molecules and atoms, but his attempts at this were unsuccessful. Maxwell made the initial headway in this direction. He showed that the atoms in a gas travel at a variety of speeds that fall within a specified range, or distribution. In fact, the speeds of the atoms in this range are such that a very small number travel either slowly or fast, with the majority of molecules traveling at intermediate speeds. This was the first time a statistical approach was used to describe the physical properties of a system. Moreover, it eliminated the daunting and, for all practical purposes, impossible task of considering all the collisions that occur in a system of gas atoms.

We saw earlier that at room temperature and pressure, a balloon filled with helium contains some 10^{23} atoms that travel at an average speed of about 2800 miles per hour, with a single atom making about a billion collisions per second with other atoms. How could one ever hope to consider all these explicitly?

Maxwell showed that we don't need to, thus opening the door for Boltzmann. Boltzmann showed that Maxwell's approach applied not only to the speed of an atom, or its kinetic energy, but also to the total energy of a system in terms of the Boltzmann distribution. This approach changed everything.

Using Boltzmann's distribution, one can calculate many properties of the system that can be observed. So while atoms and molecules are not something we can see, the theory that Boltzmann built upon their existence leads to properties we do indeed see. However, in Boltzmann's time the existence of atoms and molecules was still a very heated debate, and he was constantly defending his theory.

Today, we take it for granted that matter is made up of atoms and molecules. However, the history of the atom, as well as the theories and experiments leading up to its discovery, is a truly remarkable story and is the topic of Part III.

PART III

The Pieces: Atoms

... we find that two molecules [or atoms] of the same kind, say of hydrogen, have the same properties, though one has been compounded with carbon and buried in the earth as coal for untold ages, while the other has been "occluded" in the iron of a meteorite, and after unknown wanderings in the heavens has at last fallen into the hands of some terrestrial chemist.

– JAMES CLERK MAXWELL, SCOTTISH PHYSICIST (1831–1879)

CHAPTER 9

Speculations of Atoms

Thoughts of Existence Pave the Way for Atoms

The ancient Greek philosophers played a significant role in shaping the initial thoughts about atoms. Several of the ancient philosophers pondered and developed a theory of matter, with one even imagining the existence of a fundamental building block that made up not only all living and nonliving things, but the supernatural as well. Their thoughts were speculative and philosophical, rather than scientific in nature. And while they attempted to touch on the nature of matter and its composition, their real goal was to address something of profound concern to the ancient Greeks: the nature of permanency and change. Unfortunately, these "theories" of matter were rather short-lived. Although there was some revival during the Middle Ages and the Renaissance, they never gained any real momentum until the seventeenth century.

Permanency and Change

The Greek philosopher Heraclitus (c. 540 BC–c. 475 BC) was a native of the Greek city Ephesus on the coast of Asia Minor, or present-day Turkey. What we know of Heraclitus' works mainly comes from surviving fragments from Plato and Aristotle, where it was quoted for the sake of rebuttal, and from

Diogenes Laërtius (c. third century), a biographer of Greek philosophers, who also gives us insight into his life in general.

Heraclitus, as a pessimist, had mostly contempt for mankind. He ridiculed Homer,[101] who he claimed should have been turned out and whipped, and had nothing but scorn for several of the intellects of the time such as Pythagoras[102] (c. 560 BC–c. 480 BC) and Xenophanes[103] (c. 570 BC–c. 480 BC). Even of his fellow citizens of Ephesus he said, "[They] would do well to hang themselves, every grown man of them, and leave the city to [children]."

Heraclitus saw everything in the universe as being in a constant state of change, where nothing remained the same for even the slightest moment in time (he even said the Sun is new every day). This doctrine of flux[104] means, for example, that one would never be able to touch the same object twice; you could touch an object once, but by the time you touch it again it would have already changed its state, according to Heraclitus. But the idea that everything is changing is at odds with our everyday senses, which tells us that some things do seem to maintain a sort of permanency.

Heraclitus addresses this "appearance" of permanency with the following example: while a river gives the impression of being something permanent,

---

[101] Greek epic poet best known for his works the *Iliad* and the *Odyssey*. The dates of when Homer lived are still controversial even to this day. Herodotus provides the estimate that Homer lived 400 years before him, which would put him at about 850 BC. However, other ancient sources give dates around 1194 BC to 1184 BC, which are the dates usually preferred for the Trojan War by those who believe it to be a historical event rather than merely mythology.

[102] Greek philosopher best known for his proposition, which he made strictly by mathematical deduction, that the square of the lengths of the sides of a right triangle are equal to the square of its longest length (hypotenuse). This is still known as the Pythagorean theorem.

[103] Greek philosopher, poet, and social and religious critic.

[104] Heraclitus also believed in the mingling of opposites to produce a state of harmony. This mingling was brought about in strife, sometimes manifesting itself in war. This dualistic philosophy of Heraclitus' is something akin to the yin and yang in Chinese philosophy. Indeed, his writings and that of the Chinese classic *Tao Te Ching*, which discusses the concepts of yin and yang, supposedly written by Lao Tzu ("Old Master"), do occur around the same time. Whether one influenced the other is something we'll never know.

the water is constantly flowing, and as such the river is really in a constant state of change.[105] It's interesting to note that Heraclitus' theory of constant change isn't far from reality. Imagine an object such as a glass of water sitting on your kitchen counter. Suppose that throughout the course of the day you notice this glass of water several times over, and, of course, never is there any noticeable change in its appearance. For the most part, from a *macroscopic* perspective, the glass of water is the same. It had the same temperature, volume, and number of water molecules[106] each time you looked at it.

However, from a *microscopic* perspective things were constantly changing. The molecules of water in the glass have been and will continue to be in constant motion. To be sure, Heraclitus had no knowledge of atoms and molecules, and therefore this specific example would have never occurred to him. Although Heraclitus didn't believe in atoms, he did believe in fundamental building blocks of matter, or principal "elements."

Heraclitus believed in three principal elements: earth, air, and fire. Among these, he chose fire as the "primary element." His choice of fire as the primary element is in the same spirit as the aforementioned river. When one views fire, it's easy to imagine it as unchanging, as the flame burns without any significant signs of change. However, in reality it changes by virtue of its constant consumption of the very fuel needed to keep that same flame going. Thus, fire was the source of all other matter, according to Heraclitus: "All things are an exchange for Fire, and Fire for all things ...." Moreover, while everything in the universe is in a state of constant change, the universe itself is eternal. He says, "was ever, is now, and ever shall be an ever-living Fire ...."[107] While Heraclitus promoted the universe as being in a constant state of change, Parmenides was claiming the exact opposite.

The Greek philosopher Parmenides (c. 515 BC–c. 450 BC) was born into a wealthy and illustrious family in the Greek city Elea, on the southern

---

[105] Heraclitus said, "You cannot step twice in the same river; for fresh waters are ever flowing in upon you."

[106] There may be some loss of water molecules from evaporation.

[107] Heraclitus believed his everlasting fire to have a counterpart in the human soul. This is rather reminiscent of Hinduism, where the universal soul, known as Brahman, has a mirrored copy in all human beings, known as Atman.

coast of present-day Italy. He was the founder of the School of Elea, and his most prominent pupil was Zeno (c. 490 BC–c. 425 BC), whose works may have been the first to use the method of *reductio ad absurdum*.[108] Parmenides had a strong influence on other philosophers, particularly Plato, who spoke highly of him and even wrote a dialogue named after him. Indeed, Plato's "Theory of Ideas"[109] may have been influenced by the ideas of Parmenides. All that survives of Parmenides' work is the poem *On Nature*, which sets forth his doctrine.

In the first part of the poem, "The Way of Truth," Parmenides tell us the universe is changeless, and what we think is change is merely an illusion. Part of the disparity between Heraclitus and Parmenides arises from the fact that Heraclitus believed unconditionally in one's sense, whereas Parmenides insisted that our senses serve only to deceive us. For Parmenides, simply thinking of something gave it an existence: "Only that can really exist which can also be thought."

The essence of Parmenides' argument against change goes like this: every time you think of or speak of something, that thought or word must pertain to something that actually exists. In other words, in order for a thought to be conceivable and language to be sensible, they both require objects outside themselves. Now, since at any given moment you can think of or speak of anything, everything must exist at all times. Therefore, there can't be change since change requires something to either come into existence or pass from existence.

So, according to Parmenides if something exists, it exists for all time, and therefore it has no beginning or end, which makes it infinite in a sense. Moreover, arguing against the possibility of more than one thing being infinite (infinity is a unique state), Parmenides makes his second big conclusion: everything that exists is unified under "The One." Parmenides was a strict monist.

---

[108] It's Latin for "reduction to the absurd." It's a powerful method in mathematics and philosophy. The approach goes as this: first state the initial proposition, show that the stated proposition leads to a contradiction, and thus the initial proposition is concluded to be false as a result of the contradiction.

[109] Plato's "Theory of Ideas," also called "Theory of Forms," was surely also influenced by Socrates.

OK, so if you're having trouble following some of the logic of Parmenides, you're not alone. His original writings are hard to follow, and we are indebted to others for clarification. Besides, today we agree that Parmenides' logic was unsound. The take-home message is that Parmenides, unlike Heraclitus, believed change was merely an illusion brought into existence only as a result of our deceptive senses; the real state of the universe is a state of permanency. In addition, he believed that the universe was unified under a single existence rather than several individual ones. Finally, take a moment to realize that Parmenides' insistence on our senses being nothing but tools of deception implies the impossibility of ever realizing meaningful experimental science.

These main points of Parmenides, along with Heraclitus' insistence that the actual state of the universe was one of constant change, had an enormous impact on succeeding philosophers. This is especially true of those who developed theories of matter. The main goal of these ancient theorists was to unify, in some sort of way, the disparate doctrines of Parmenides and Heraclitus, rather than to construct a complete physical theory of matter. Nonetheless, out of their initial efforts we find the beginnings of atomic theory, with one in particular bearing a striking resemblance to modern atomic theory.

Ancient Theories of Matter

The first "atomic theories" focused on a "primary element" responsible for creating all other matter. Heraclitus said it was fire, Thales of Miletus (c. 624 BC–c. 546 BC) said it was water, Anaximenes (c. 585 BC–c. 528 BC) thought it was air,[110] and Empedocles finally unified these, declaring

---

[110] Anaximenes' theory of air as the primary matter or source of all matter is rather interesting. By implementing the physical mechanisms, very familiar to us today, of condensation (compression of air to make it denser) and rarefaction (expansion of air to make it less dense), he developed an explanation of how the other "elements" occur. Simply put, he said air undergoes condensation to form water, and then condensed further it becomes earth. However, when air undergoes rarefaction it's thinned (less dense) and becomes fire. In this way, he provides a basis for these four elements along with a theory of change to rationalize their transformations.

there to be four elements: air, earth, fire, and water. Later, Aristotle adopted Empedocles' four elements, and so it remained until about the seventeenth century.

Born in Acragas, in present-day Sicily, Empedocles (c. 492 BC–c. 432 BC) was an interesting character who has been described as a philosopher, prophet, healer, democratic politician, mystic, charlatan, fraud, and scientist. His main contribution to the physical sciences was his four-element theory.

These fundamental elements, which he called "roots," combined in varying amounts to form all other matter: plants, animals, humans, rocks – everything. And while the elements mixed together to form other things, they still maintained their own individual characteristics. Indeed, Empedocles envisioned the four elements as changeless, eternal, and indestructible. Empedocles believed in two eternal metaphysical forces: Love and Strife. Love was responsible for bringing the elements together in the process of creation, whereas the opposing force of Strife was responsible for the separation of the elements, ultimately leading to the process of decay. The cosmic battle between Love and Strife represented the natural cycle of change in the universe; Love built things up and Strife tore them back down, and they struggled against each other, each one trying to gain dominance over the other.

In Empedocles' theory, we clearly see the concept of a changing universe similar to that described by Heraclitus, although where Heraclitus only believed in Strife, Empedocles softened his theory with the addition of Love as its cosmic counterpart. Perhaps less clear is that Empedocles also embraced a bit of the doctrine of Parmenides. While he didn't believe in a changeless universe, as Parmenides' monistic dogma had demanded, he did attribute changelessness to his fundamental elements. To be sure, this was a deliberate attempt to reconcile the opposing doctrines of Heraclitus and Parmenides, and he was not the only one to do so. Nonetheless, he abandoned Parmenides' monistic view in favor of a pluralistic one governed by his four elements (roots), two forces, and the ensuing comingling thereof. A contemporary of Empedocles and fellow early atomic theorist was Anaxagoras.

Anaxagoras (c. 500 BC–c. 428 BC) was born in the town of Clazomenae in Ionia, located in present-day Turkey. He was the first to bring philosophy to Athens (most likely convinced to come by Pericles (c. 495 BC–c. 429 BC) who became his student) and spent thirty years there but eventually left. It seems his teachings on the Earth, the Sun, and the Moon[111] might have gotten him into a bit of trouble (apparently Galileo wasn't the first to suffer this fate). Specifically, he was in violation of a law permitting the impeachment of anyone who didn't practice religion and taught theories about the heavenly bodies. He was charged for impiety. Rather than stay and face his sentence, which was execution, with the assistance of Pericles he left Athens for Lampsacus (in Asia Minor), where he remained for the rest of his life.

Like Empedocles, Anaxagoras attempted to meet the challenge of Parmenides' demand of a changeless world while accounting for the apparent change we experience in everyday life. Whereas Empedocles singled out air, earth, fire, and water as the fundamental building blocks of everything

---

[111] In *The Refutation of All Heresies*, Hippolytus discusses Anaxagoras' beliefs on the Earth, the Sun, and the Moon:

And that the Sun and Moon and all the stars are fiery stones, that were rolled round by the rotation of the atmosphere. And that beneath the stars are Sun and Moon, and certain invisible bodies that are carried along with us; and that we have no perception of the heat of the stars, both on account of their being so far away, and on account of their distance from the Earth; and further, they are not to the same degree hot as the Sun, on account of their occupying a colder situation. And that the Moon, being lower than the Sun, is nearer us. And that the Sun surpasses the Peloponnesus in size. And that the Moon has not light of its own, but from the Sun. But that the revolution of the stars takes place under the Earth. And that the Moon is eclipsed when the Earth is interposed, and occasionally also those [stars] that are underneath the Moon. And that the Sun [is eclipsed] when, at the beginning of the month, the Moon is interposed. And that the solstices are caused by both Sun and Moon being repulsed by the air. And that the Moon is often turned, by its not being able to make head against the cold. This person [Anaxagoras] was the first to frame definitions regarding eclipses and illuminations. And he affirmed that the Moon is earthy [similar to the Earth], and has in it plains and ravines. And that the milky way is a reflection of the light of the stars which do not derive their radiance from the Sun ....

in existence, Anaxagoras, seeing no reason for such discrimination,[112] declared that everything contains a bit of everything else.

For Anaxagoras, things such as bone, skin, and hair were just as real as Empedocles' fundamental elements of air, earth, fire, and water, and as such there's no reason one would choose some in favor of others. So he decided not to choose, but instead included a "portion of everything in everything." While it's not known for sure, it might be that he arrived at his theory of matter from his rather insightful studies in nutrition. Anaxagoras noted that food provided nourishment for animals that, in turn, caused certain things to occur, such as their hair to grow and their skin to heal. He concluded that food must then have the constituents of hair and skin already in it to be able to convey these effects.

Moreover, Anaxagoras considered matter to be infinitely divisible. Thus, if one cut a piece of hair again and again, it would still contain the essence of hair. He says, "For of the small there is no smallest, but always a smaller (for what is cannot not be). But also of the large there is always a larger, and it is equal in amount to the small. But in relation to itself, each is both large and small."[113]

---

[112] According to Aristotle, Anaxagoras felt that Empedocles unjustly singled out certain substances as primary and others as secondary. Moreover, Anaxagoras probably had another problem with Empedocles' theory. According to Empedocles, everything could be divided into the four fundamental elements. Thus, something, such as an orange, when it decays simply does so into air, earth, fire, and water. Literally speaking though, this means the thing that we once called an orange is now gone, and all that remains (somewhere) are those four elements. The problem is that it seems the orange has disappeared; or, rather, it does not exist anymore. In other words, "something" has become "nothing," which was a contradiction of Parmenides' doctrine of the impossibility of such a scenario. Therefore, Anaxagoras may have been trying to maintain this portion of Parmenides' doctrine when he declared "a portion of everything in everything." However, for Empedocles the comingling and separation of the four elements under the forces of Love and Strife respectively was the reality, whereas the "orange" was an illusion or artificial assignment given by human interpretation. Thus, nothing is lost from existence since the four elements constituting the orange remain when the orange is gone; they merely went from a state of comingling to one of separation in accordance with Empedocles' doctrine.

[113] It has been argued from this surviving fragment of Anaxagoras' work that he had an unprecedented understanding of the mathematical concept of infinity.

However, despite this, Anaxagoras considered these components of matter, often referred to as "seeds," or "stuffs," as eternal and indestructible, albeit more loosely so than envisioned by Empedocles. This still raises the question that if everything contains everything else (in varying proportions nonetheless), then what is it that makes something what it is? To this Anaxagoras replies: "each single thing is and was most plainly those things of which it contains most." In other words, something is what it is because it contains most of that "stuff." More precisely, something is what it appears to be *macroscopically* because it contains most of that "stuff" *microscopically*.

So we see in both the theories of Empedocles and Anaxagoras the attempt at producing intelligible theories of matter. Each tried to combine ideas that would account for the changing world that we all experience, while still allowing for certain components to remain fundamental, and as such changeless. In effect, each was trying to include simultaneously, in his unique way, the dogma imparted by Heraclitus and Parmenides.

Today, their ideas may sound strange and metaphysical to us, yet one can find the similarity between them and current atomic theory. This is perhaps best exemplified in the theory put forth by Democritus, who is undoubtedly the most important ancient Greek atomic theorist.

Democritus' Atom

Democritus (c. 460 BC–c. 370 BC) was a native of Abdera in Thrace, located in present-day Greece. He traveled widely, perhaps spending time in Egypt and Persia. He also spent time in Athens: "I went to Athens and no one knew me."

Indeed, it seems that in Athens Democritus never really fit into the intellectual elite, and his philosophy was ignored for some time. Nonetheless, his wealth of knowledge and exactness of thinking give him a much-deserved place in the history of philosophy. By our present-day standards, he was perhaps the most successful of the ancient Greek philosophers with regards to the remarkable accuracy of his ideas. For example, he considered the Milky Way to be a collection of tiny stars, and the Moon to be very much like the Earth in that it contained mountains and valleys. Regardless, we know him primarily for his atomic theory.

Democritus was a student of Leucippus (fifth century BC), who had an atomic theory of his own. In fact, it's hard to untangle the atomic theories of Democritus and Leucippus. This is mostly because we know very little about Leucippus, and it has been speculated that he never actually existed, although this seems unlikely since Aristotle and Theophrastus (c. 371 BC– c. 287 BC) mentioned his atomic theory explicitly. It seems more likely that Leucippus set in place some of the fundamentals, and that Democritus built upon them, thereby extending the overall theory.

Democritus considers everything in the universe – including the human mind and soul,[114] and even the gods – to be made up of *atomos*, which is Greek for *indivisible* and from which we get the word *atom*. Indeed, Democritus considered these atoms to be indivisible (contrast this with Anaxagoras, who considered his fundamental pieces to be infinitely divisible). He imagined atoms to occur in a variety of different shapes and sizes, which were responsible for the properties found in the objects they made up. Moreover, he considered atoms to be changeless, eternal, and indestructible, similar to the way Empedocles envisioned his four fundamental elements.

Democritus saw material objects as existing in a temporary state, being created or destroyed as atoms come together or fall apart under the influence of natural forces; all that remains, then, are the atoms constituting those material objects. This is not unlike Empedocles' view, where he imagined the four elements giving rise to material objects under the influence of the forces Love and Strife. In addition, Democritus also gave motion to his atoms.

Democritus imagined atoms as always in motion, undergoing collision after collision with each other as they moved around. Moreover, this motion was a fundamental property and, like the atoms themselves,

---

[114] Democritus believed the human soul was made of special atoms that were very fine and round, allowing them to penetrate throughout the body, which was necessary for all its essential functions.

was eternal and indestructible, although changeable under certain circumstances.[115]

In order for atoms to be in motion, there must be a space for them to move, and thus Democritus invented the *void*. According to Democritus, atoms move in the void with a constant random motion (he likened the movement of atoms to the dust particles one sees dancing around in the sunlight when there's no breeze). This is much like how we imagine them doing so today, as described by modern-day kinetic theory.

Recall that in Parmenides' philosophy, material things have existence because we are able to think of them. He also considers it impossible to think of nothing, and therefore it can't exist. Thus, Democritus' void may seem to be in blatant disregard of this tenet, as for all practical purposes it seems to be nothing. However, Democritus saw the void as something: a place independent of the atoms for the atoms to reside and move around in. The real problem is that Parmenides could only imagine material objects as something, whereas Democritus was able to imagine both a material object (the atom) and the space it lived in as being something. Democritus makes his point clear: "Nothing exists except atoms and empty space; everything else is opinion."

Democritus made concessions to both Parmenides and Heraclitus, just as Empedocles and Anaxagoras did, by imagining a universe consisting of an infinite number of changeless, eternal, and indestructible atoms, always engaged in random collisions with each other, and capable of comingling to form material objects as we know them.

Aside from its remarkable similarity with modern-day atomic theory, Democritus' atomic theory is redeeming in itself for the very fact that it offers a "mechanical explanation" for matter: matter is made of atoms that move in a void and undergo collisions (where preceding collisions are determined by previous ones) that are governed by certain physical laws of nature.

He invokes no divine intervention in this atomic process, but quite simply he maintains that atoms have always been and always will be in

---

[115] It's not exactly clear how Democritus envisioned a change in motion occurring, although apparently it could result from an applied pressure.

motion, and that physical laws describe this motion. The beauty of such a construct is that it lends itself to a scientific description. That is to say, one can hope to develop a mathematical theory describing the physical laws and then proceed to perform experiments to test this theory.

Obviously, neither the needed mathematics nor the experimental procedures were available to Democritus. Additionally, Democritus' theory suffered another blow – namely, Aristotle,[116] who stunted the development of Democritus' work. On several accounts he mentions[117] Democritus' atomic theory explicitly, only to attack it. Ironically, it's in this way that we learn much, perhaps the majority, of what we know of Democritus' atomic theory.

Why Aristotle?

Aristotle (c. 384 BC–c. 322 BC) was born in Stagira, Greece. His father was the personal physician to the King of Macedonia, a position he inherited. Aristotle studied with Plato (c. 427 BC–c. 347 BC) in Athens beginning at the age of eighteen and remained there for nearly twenty years until Plato's death. In 343 BC, Aristotle became the tutor to Alexander the Great, who was then thirteen, and continued until he was sixteen, when Alexander's father made him regent at Pella.[118]

Aristotle's writings provided the first comprehensive system of Western philosophy covering topics in politics, ethics, logic, metaphysics, and science. There was hardly an area he didn't write about.[119] Believing all of human knowledge couldn't fall under a single category, Aristotle was the first to divide it into categories. Here, we are interested in Aristotle's theory of matter and form.

---

[116] As did Plato, but not to the same extent Aristotle did.

[117] We learn much of Democritus' atomic theory through Aristotle's attacks on his work, particularly in *On Generation* and *Corruption* and *On the Heavens*.

[118] This meant he was in charge while his father (Phillip II) was away. Indeed, Alexander put down an uprising in the region of Thrace and established the town Alexandropolis.

[119] As Galileo apparently noted in frustration, having had to learn Aristotle's works while attending the University of Pisa.

Just as others did, Aristotle sought to rise to Parmenides' challenge of permanency while maintaining room for change in the world, as Heraclitus had demanded; his theory of matter and form is an attempt at this reconciliation.[120] According to Aristotle, objects as we know them comprise two parts: "matter" and "form." The form gives a particular arrangement to matter, and it's by virtue of the form that we identify an object as a "thing"; to know a thing is to have knowledge of its form.

For example, imagine that a sculptor starts with a lump of clay and proceeds to mold it into the shape of a dog. Here the clay is the matter, and the shape of a dog conferred to the clay by the sculptor is the form. Now, imagine the sculptor begins again, transforming the piece of clay, which once held the shape of a dog, into something else, perhaps a cat this time. Clearly, the matter is still the clay, but now the form has changed from that of a dog to a cat. However, the sculptor does not create the form; it was always there. Instead, the sculptor's efforts merely brought the form and the matter together. According to Aristotle, change results from a change in the form of matter.

Moreover, Aristotle describes such a process as being governed by four causes: material, formal, efficient, and final. These are the axioms governing the way a material object comes to be, and you can think of them in terms of these questions: What's the material the object is made of? What's the object? How was the object built? What's the purpose of the object? The most important of these is the last one, known as the final cause. Indeed, if there were a central tenet underlying Aristotle's philosophy, it would be the question posed by the final cause.[121]

It's the final cause that provides a certain goal for the matter as it moves through its various forms. For the most part, it's the final cause that provides

---

[120]   It's also an attempt to improve upon what he saw as weaknesses in Plato's Theory of Ideas (Forms).

[121]   When one draws a conclusion based on supposing a purpose, it's called a teleological explanation. In contrast, when one draws a conclusion based on possible causes, they are implementing a "mechanical explanation." Historically, we have found that it's the latter that facilitates scientific knowledge, not the former. Aristotle used teleological explanations, whereas Democritus used mechanical explanations.

a sense of permanency throughout the overall process. Thus, Aristotle's doctrine of matter and form attempts to unify the seemingly disparate ideas of change and permanency. Related to Aristotle's theory of form and matter are the concepts of "potentiality" and "actuality." Again, consider the sculptor and the clay. When the clay was merely a lump on the sculptor's workbench, it had only the potentiality to take the form of a dog or a cat, among other things. But when the clay acquired form through the sculptor's efforts, it increased its actuality. Thus, the more form something has, the greater its actuality is. Aristotle worked these principles into his theology as well, where his version of God is depicted as perfection consisting of pure form and actuality.[122]

Aristotle's works were rediscovered after the fall of the Roman Empire by the Arab civilization ruling the region spanning from Persia to Spain. Among this group of Arabs were Muslim and Jewish scholars, who translated the works of Aristotle (and virtually every important work in Greek culture, as well as Persian and Indian culture) into Arabic. These translated works were then acquired by medieval Christians, who by 1100 began to gain control over this Arab civilization in regions such as Toledo, Spain, and Lisbon, Portugal.

The Muslim and Jewish scholars included addendums to the original works. Thus, not only did they translate the original works from Greek to Arabic, they also completed ideas left unfinished by the ancient Greeks, thus enhancing the original works. The timing couldn't have been better for Christian scholars, because by the mid-twelfth century they were already beginning to wonder about the relationship between God and, well, everything else. It was Aristotle who provided them with the insight they were looking for – that is, once they had all his works translated from Arabic to Latin.[123]

---

[122] For Aristotle, God is pure form, existing without matter, and pure actuality, and therefore is not changing since he is perfection. In fact, all things are striving towards a likeness with God as their final goal. This means that God is the ultimate final cause for all motion and change. Aristotle's theology is one where material things are "evolving" towards a likeness with the perfection of God.

[123] Translation centers were set up by the Christians to have the works translated from Arabic to Latin (and some directly from Greek to Latin). This was an amazingly

There were probably several reasons Christian scholars favored Aristotle over the other ancient Greek philosophers. For one thing, he provided a very complete system of philosophy, having commented on just about everything. His writings were written in a very academic manner while still being very tractable to a general audience, having just enough common sense mixed in. Aristotle's common sense, in part, came from the fact that he was very much an empiricist – whereas Democritus was more theoretical in thought, Aristotle was more observational; he observed nature and believed we could acquire useful information from the world in this way. Finally, Aristotle's vision of God, although not that of a Christian God, evidently provided enough of a starting point to be integrated into a new version of Christianity of the time, thanks mostly to the likes of St. Thomas Aquinas (1225–1274).

Once successfully integrated into Christianity (and with the early European universities being tied to the Christian Church), Aristotle became the authority on just about everything, in particular science, up until about the seventeenth century. So the works of Democritus really didn't have a chance to flourish for these reasons and a few others.[124] Nonetheless, the seventeenth century would soon change all that as scientists sought to understand the world in a more systematic (mechanistic, or mechanical) way with the new tools available to them in the rapidly changing areas of physics and mathematics.

---

cooperative effort where Muslims, Jews, and Christians worked together on this common goal and mainly coexisted nicely.

[124] We need to remember that Democritus' atomic theory attempted to answer the fundamental nature of change of permanency – this was the big question on the minds of the ancient Greeks at that time. If the question would have been more simply, "What are the fundamental building blocks of matter?" then perhaps his atomic theory would have caught on a bit more. Additionally, by the Middle Ages Aristotle was considered the expert on most things, as we have discussed, and offered harsh criticism of Democritus' work. Finally, as with all the pre-Socratic philosophers, Democritus' work only survives in the form of fragments and hearsay from sometimes unreliable sources.

CHAPTER 10

Two New Philosophies

Rational Versus Spiritual View of Nature

The seventeenth century was revolutionary for mathematics and the physical sciences. It saw the emergence of numerous powerful techniques in mathematics, which also provided new approaches never before available, to the solution of physical problems. These new tools complemented the new *mechanical philosophy*, which proposed a systematic view of the world, freed from an intervening god, that ran according to fundamental laws set down by nature and comprehensible to man.

A competing, or perhaps complementary, "philosophy" to the mechanical philosophy was *alchemy*. With its metaphysical and spiritual doctrine, it served to both hinder and promote the beginnings of experimental chemistry. As alchemy faded away, experimental chemistry continued to evolve into a mature science that acknowledged the atom as conceptually useful, but was still reluctant to fully embrace it as physical reality.

Aristotle's End

Aristotle's doctrine on matter, like everything else he said, was gospel. However, it was failing to hold up under ever-increasing criticism. Recall that Aristotle thought an object consisted of two parts, matter and form, which coexisted within an object, making it what it is. They not only gave shape to the object but also were responsible for all the physical and chemical properties of the object.

The problem here is that regardless of the object and its characteristics, one is forced to conclude that its properties arise simply because they're innate to the object via its form. So, why is the sky blue? Because "blueness" is an innate part of its form. Why is gold shiny? Because "shininess" is part of its form. See the problem here? We end up learning nothing about the underlying principles or mechanisms that really govern the properties of the object in question.

Democritus' atomic theory suffered greatly from the criticisms of Aristotle, but it's not as if it remained in total obscurity. On the contrary, in the Middle Ages[125] both the atomic theories of Democritus and Epicurus (c. 341 BC–c. 270 BC)[126] had been known and were of some interest even during this period.

Moreover, a resurgence occurred around the time of the Italian Renaissance,[127] thanks to Poggio Bracciolini's[128] (1380–1459) discovery of a single, barely surviving manuscript in 1417 of the great work *De Rerum Natura* (*On the Nature of Things*) by Roman Poet Lucretius (c. 99 BC–c. 55 BC).

---

[125] The period of European history that occurred from the fifth (following the fall of the Roman Empire in 476 AD) to the fifteenth century.

[126] For the most part, Epicurus' atomic theory follows Democritus' with one significant refinement. According to Epicurus, atoms in the void move in undisturbed downward paths that are parallel to each other. However, at random times some atoms will spontaneously "swerve" to the side. According to Epicurus, the resulting collisions allow the atoms to come together and form the material objects we see in the world, as well as being responsible for other natural phenomena. For Democritus, subsequent atomic collisions resulted as a consequence of previous ones. Thus, atomic motions – and things in general – were viewed by Democritus as deterministic. Epicurus couldn't buy into a totally deterministic view of the world, which would eliminate free will altogether, and therefore added "the swerve" to his atomic theory to include enough randomness to allow for the concept of free will.

[127] The period of Italian history that occurred from the end of the thirteenth century to about 1600.

[128] Bracciolini was a humanist of the early Italian Renaissance.

Lucretius was a contemporary of Julius Caesar (100 BC–44 BC) and Cicero (106 BC–43 BC). He believed that atoms were extremely small particles of matter having the same properties as the bulk matter they were a part of. His view was widely held until around the nineteenth century, when it was realized that atoms have some unique properties separate from the bulk, such as electrical charge. Today, we know that *bulk properties* result from the collection of the large number of atoms making up the matter of a given object, and thus the properties of a single atom don't directly translate to those of the bulk.

Lucretius' poem, which documented and embellished the works of Democritus, Leucippus, and Epicurus, was very popular in the fifteenth and sixteenth centuries as both a literary and philosophical work. The work of Diogenes Laërtius, which had been mostly ignored in the Middle Ages, regained popularity, especially the tenth book of his *Lives of the Philosophers*. This work was of particular importance for the development of atomic theories, containing three letters by Epicurus (*Letter to Herodotus*[129] being the most important) in addition to an account of Epicurus' life.

Throughout the sixteenth century, the concept of atoms was of interest in physical theories and philosophical doctrines. By the middle of the sixteenth century, the idea that matter may indeed be made up of atoms was gaining widespread interest. A quote from Nicolaus Copernicus[130] expresses this sentiment: "As with those tiny and indivisible bodies called atoms which, though they are not perceivable by themselves and do not when taken two or several together immediately form a visible body, yet may be multiplied until they join to form finally a great mass ...." However, it would be the seventeenth century that would see a full resurgence of the ancient theories of matter.

The seventeenth century saw dramatic advances in mathematics and physics, which provided scientists with the tools to take a closer look at

---

[129] Herodotus was an ancient Greek historian (c. 484 BC–c. 425 BC).

[130] Nicolaus Copernicus was a Renaissance astronomer who was the first to argue that the Sun was the center of the universe, which went against the established belief of the time that the Earth was at the center. He detailed his work in *On the Revolutions of the Celestial Spheres*, which, because of fear of persecution from the Church, he waited to publish right before his death in 1543.

many unanswered physical problems, among them being the real nature of matter, which brought the early atomic theories, especially Democritus', to the foreground. Democritus' atomic theory, even with all its shortcomings, was still redeeming in that it provided a "mechanical" explanation of matter: matter consists of atoms in motion governed by certain (then yet to be determined) physical laws, and the atoms themselves give rise to the physical and chemical properties of the object in question.

To seventeenth-century scientists, this picture of matter was more appealing than Aristotle's doctrine of matter and form (which eventually suffered ridicule) because it identified fundamental features that one could hope to exploit experimentally and theoretically to gain a deeper understanding of the fundamental workings of nature. Thus, nature was viewed as a sort of grand "machine" comprising certain "parts" that moved only according to physical laws (as opposed to driven by some all-powerful being in the universe). In fact, a favorite metaphor of the seventeenth century was the comparison of the workings of nature to that of a mechanical clock, as so aptly captured by Robert Boyle (1627–1691):

> The several pieces making up that curious engine are so framed and adapted, and are put into such a motion, that though the numerous wheels, and other parts of it, move several ways, and that without any thing either of knowledge or design; yet each part performs its part in order to the various ends, for which it was contrived, as regularly and uniformly as if it knew and were concerned to do its duty.

This new view of nature as working much like a machine, or clock, was an essential step towards freeing the seventeenth-century scientist from Aristotle's doctrine. It stood for the resounding belief that one could systematically understand the phenomena of nature rather than having to resort to vague explanations of underlying purpose, innate qualities, or occult powers as governing these processes. This new way of viewing nature was known as the *mechanical philosophy*.

The World Machine

In a very telling passage from *The World* (written between 1629 and 1633),[131] René Descartes (1596–1650) makes it clear that he is a proponent of the mechanical philosophy and considers the Aristotelian concept of form and its associated qualities incapable of explaining physical phenomena:

> If you find it strange that I make no use of the qualities one calls heat, cold, moistness, and dryness ... as the philosophers [of the schools] do, I tell you that these qualities appear to me to be in need of explanation, and if I am not mistaken, not only these four qualities, but also all the others, and even all of the forms of inanimate bodies can be explained without having to assume anything else for this in their matter but motion, size, shape, and the arrangement of their parts.

Undoubtedly, like many of his contemporaries, Descartes' support for the mechanical philosophy was largely motivated by his desire to disprove Aristotelian philosophy. Descartes made reference to machines in models he used to explain the behavior of the world. In fact, for him there was really no fundamental difference between man-made machines and nature's machines: "There is no difference between the machines built by artisans and the diverse bodies that nature alone composes."

The only exception for Descartes was that the parts making up man-made machines needed to be large enough for one to see, whereas nature's "parts" may be so small as to be invisible to us. As we learned in Part I,

---

[131] In 1634, Descartes was finally ready to publish *The World*, but then found out that the Inquisition at Rome had condemned Galileo for the teachings in his *Dialogue Concerning the Two Chief World Systems*. Now realizing that *The World* contained material that would be seen as controversial by the Church, he feared suffering the same fate as Galileo, and in 1634 wrote to a friend that he wouldn't publish. In 1662, the last chapter of *The World* was published separately as *On Man*, with the rest published in 1664; in 1677 it was published in its entirety.

Descartes asserted that "movement" in nature was conserved.[132] So for Descartes, while God might have put everything in motion at one point, such as the planets, thereafter he is not needed to keep the world machine[133] going.

Thus, the role of God was the architect and creator of the world machine, and this machine, once set in motion, would run forever without God's interference. The world machine picture appealed to many scientists of the seventeenth century because now nature should be describable according to a set of well-defined mathematical principles and physical laws – it's a machine, after all.

Perhaps no other person did more to lay the foundation and initiate the success of this machinery than Newton. Yet, in one of the greatest ironies of history, he essentially rejected his own design. In 1687, Isaac Newton

---

[132] Today, we say that linear momentum and angular momentum are independently conserved. We learned about linear momentum in Part I, and you'll recall that it's the mass of an object multiplied by its velocity. (The velocity of an object defines both the speed and the specific direction of motion; it's a vector.) An excellent example of the conservation of linear momentum is seen in the game of pool. Upon breaking, when the cue ball collides with the rest of the balls, it transfers linear momentum to them and consequently loses linear momentum itself; overall then, the linear momentum is conserved. Angular momentum is also conserved. You can think of angular momentum as the rotational counterpart of linear momentum. Holding a cup of coffee in your hand and then slowly rotating your body in one direction is a cool example of the conservation of angular momentum. If you rotate to the left, the coffee will spin to the right in the cup and vice versa. The system, being you and the cup of coffee, is conserving angular momentum so that the total amount is zero, which was the amount that was present to begin with when you weren't moving. That's conservation: the beginning and ending amounts are the same (i.e., zero).

[133] Descartes did insist that the world was very much like a machine, even with respect to the physical workings of the human body and animals in general. However, he did make a clear distinction when it came to the human soul. For Descartes, if mental attention of any sort was involved in an act then it required the involvement of a soul. Moreover, this soul and the complex machinery of the human body were connected according to God's will. It's interesting to note some of the things that Descartes felt didn't require mental attention, things that we often attribute a psychological component to, such as perception, memory, sensations like fear, or actions such as waking or running.

(1643–1727) wrote *Principia*, which provided (among other things) three fundamental laws describing the motion of objects and the role of forces therein.

Newton's First Law states:

> An object at rest will remain at rest (zero speed) and an object in motion will continue its motion at constant speed and in a straight line, unless an applied force alters either.

In other words, if an object is not moving, it won't suddenly start moving without an applied force – a push or a pull – putting it into motion.

Moreover, an object that is moving at a given speed in a straight line will continue to do so unless, once again, a force intervenes to cause a change. Since not moving is simply a special case of constant speed (the speed is constantly zero), we can more succinctly say that an object's natural tendency is to maintain its speed and direction in a straight line. This is called *inertia* and it's a fundamental property of matter.

Galileo and Descartes both discussed the concept of inertia, and Newton's First Law formalizes it.[134] While this may not seem like a big deal these days, it was at the turn of the seventeenth century when Galileo first suggested it. Before then, most people thought (thanks to Aristotle) that the natural state of an object was to remain at rest. The idea was that a constant force was actually needed to keep an object in motion, otherwise it would be at rest. In fact, this does seem to be more consistent with everyday experience.

For example, when driving your car down the road, a constant force (giving it gas) is needed to keep it moving. However, an outside force is acting on your car, which is the friction between the road and the tires. It's this force that causes your car to slow down, otherwise it would have kept moving as a result of its inertia.

---

[134] Newton gives Galileo due credit but ignores Descartes who, in his *Principles of Philosophy* (1644), states (as his first law of motion): "that each thing, as far as is in its power, always remains in the same state; and that consequently, when it is once moved, it always continues to move."

Galileo's experiments with rolling objects down inclines (as we learned about in Part I) led him to the concept of inertia. Both the objects and the incline were made of hard materials and as a result, the friction between them was very small.[135] Thus, as Galileo observed the object traveling a substantial distance after rolling down the incline, he imagined the idealized situation where friction was completely absent and concluded that under these circumstances the object would have continued forever under its inertia. It was these types of "thought experiments" that led Galileo to many of his insightful conclusions.

Newton's Second Law describes the effect an outside force acting on an object has on its motion:

> The acceleration of an object is *directly* proportional to the applied force and *inversely* proportional to the mass of the object.

In other words, if the applied force F produces acceleration a of an object with mass m, then Newton's Second Law is written mathematically as:

$$F = ma$$

where the force and the acceleration are *vectors* (they have magnitude and direction) and the mass is a *scalar* (has magnitude only). In terms of the acceleration, we simply rearrange and get:

$$a = \frac{F}{m}$$

So, for a given mass, if the applied force is increased by (let's say) twofold, clearly the acceleration will be increased by twofold as well, and if the applied force is increased by threefold, the acceleration will be increased by threefold ... and so on. Acceleration of an object is *directly* proportional to applied force. Now for a given applied force, if the mass is increased by

---

[135] A good example of a low-friction scenario would be an object moving over an area of ice. Anyone who has hit a patch of ice on the road in the winter can attest to the fact that the small amount of friction can cause the car to slide.

twofold, we see that the acceleration is decreased by twofold, and if the mass is increased by threefold, the acceleration will be decreased by threefold … and so on. Acceleration is *inversely* proportional to the mass of the object. Newton's Second Law also provided the first correct definition of what an applied force actually is – previously, force was confused with momentum or energy (as discussed in Part I).

Finally, there's *Newton's Third Law*:

Two objects acting upon each other exert a pair of forces that are equal in magnitude but opposite in direction.

Sometimes we refer to this pair of forces as the *action* force and the *reaction* force. We can then say that for every *action* there's an equal and opposite *reaction*. Newton's Third Law means there's no such thing as a "lone force" – all forces in the universe occur in action–reaction pairs, with each force (of a given pair) equal in amount, but directed in opposite directions.

For example, a book sitting on a table exerts a downward force on the table due to its weight and,[136] in turn, the table exerts a force back on the book, which is equal in magnitude and directed upward. Some examples seem a little less obvious. Consider a bug that hit your windshield while you were driving. Even though the bug clearly made out much worse than the car, the bug felt exactly the same force the car did.

Newton's Laws of Motion essentially provided reassurance that the world ran much like a well-oiled machine. However, Newton didn't quite see it this way. In *Opticks* he writes:

For while comets move in very eccentric orbs in all manner of positions, blind fate could never make all the planets move one and the same way in orbs concentric, some inconsiderable irregularities excepted which may have arisen from the mutual actions of comets

---

[136] What we call weight is the force of gravity from the Earth "pulling" downward on the book. Therefore, according to Newton's Third Law, this "pull" from the Earth on the book is countered with a "pull" right back from the book on the Earth, which is equal in amount but opposite in direction.

and planets upon one another, and which will be apt to increase, till this system wants a reformation.

Newton simply couldn't believe in the "extreme" version of the world machine where God was simply the architect and creator, as Descartes (and many others) had imagined. For example, he feared that it was possible for a runaway comet to cause the planets to stray from their orbits, therefore requiring the hand of God to correct this little astronomical mishap.

This may seem contradictory to Newton's Laws of Motion, but Newton didn't see a conflict. He viewed his laws as applicable within the time frame between when God put things into motion up until they went horribly awry and needed God's help. Not surprisingly, Newton's rejection (or rather, scaled-down version of the mechanical philosophy) brought on criticism, in particular from his lifelong rival Leibniz.

In a letter written in 1715, which was soon made public, Leibniz writes:

Sir Isaac Newton and his followers have also a very odd opinion concerning the work of God. According to their doctrine, God almighty wants to wind up his watch from time to time: otherwise it would cease to move. He had not, it seems, sufficient foresight to make it a perpetual motion. Nay, the machine of God's making, is so imperfect, according to these gentlemen, that he is obliged to clean it now and then ... and even to mend it, as a clockmaker mends his work ....

However, Newton, who now dictated his response to Leibniz through Samuel Clarke (1675–1729), felt that God's role was not only as architect and creator but also as preserver. Thus, his needed intervention from time to time was not a reflection of an imperfect design but rather "the true glory of his workmanship, that nothing is done without his continual government and inspection."

So while Descartes, Leibniz, Boyle, and many others saw the mechanical philosophy and the world machine as complementary to God's existence, Newton insisted that it was not nothing short of atheism, which was something he simply couldn't accept.

As mathematicians and scientists in the eighteenth and early nineteenth centuries – in particular Pierre-Simon Laplace (1749–1827), Leonhard Euler (1707–1783), Joseph-Louis Lagrange (1736–1813), and Siméon Denis Poisson (1781–1840) – further developed Newton's Laws of Motion, the belief that God's will was necessary to keep everything in balance disappeared. Evidently, Emperor Napoleon I commented on Laplace's *Treatise on Celestial Mechanics*,[137] saying that although it dealt with the universe, it made no mention of its creator. Laplace smugly replied: "I had no need of that hypothesis."

Amused by his response, Napoleon shared the story with Lagrange, who commented: "Ah, it is a fine hypothesis; it explains many things."

Newton seemed to be in the minority as to the acceptance of a mechanical philosophy free from the interfering hand of God. Perhaps this, in part, led him to explore other philosophies, including alchemy.

Lead into Gold

While the seventeenth century saw the beginnings of the mechanical philosophy and its growing popularity, there was already another "philosophy" around: alchemy. In fact, by then alchemy had been around for a very long time in one form or another.

One of the earliest mentions of alchemy is found in a Chinese imperial edict issued by Emperor Jing (Ching) in 144 BC, which established that those who made counterfeit gold, such as coiners and alchemists, were to be punished by public execution. Apparently, the previous emperor, Emperor Wen, had allowed the alchemists to create their gold. This was a problem for Emperor Jing. The problem, as he saw it, was that the alchemists were creating their fake gold, losing their money (not to mention wasting their time), and consequently driven to a life of crime, which was something Emperor Jing couldn't have.

---

[137] A set of five volumes that appeared from 1799 to 1805 with installments appearing from 1823 to 1825.

Alchemy's early history also shows up around 200 BC with *Physica et Mystica*, written by Bolos Democritos[138] (a Greek living in Egypt). Nonetheless, the main line of alchemy's development began in Hellenistic Egypt[139] in towns on the Nile Delta, particularly in Alexandria.[140] The focus of Hellenistic alchemy was on the transmutation of matter, in particular the production of gold and silver from the "base metals." This is in contrast to Chinese alchemy, which sought elixirs of longevity, immortality, and the perfection of the human soul.

Chinese alchemy was closely tied to Taoism, dating back to c. 300 BC. A similar theme was present in Indian alchemy where, perhaps as early as the eighth century BC, the Sanskrit Atharva Veda describes the use of gold as a means of preserving life. The post–eighth century tantric-Hatha yoga texts contain mystic overtones similar to those seen in Chinese sources. Indeed, mysticism was embodied in most types of alchemy.

The mysticism surrounding alchemy has often led to its condemnation as a contributor to early experimental chemistry. Nonetheless, one can reasonably argue that certain parts of alchemy did survive and were subsequently refined and expanded upon, leading to the rigorous science of

---

[138] Bolos Democritos' works were translated by Muslim Arabs in a similar fashion as discussed previously for Aristotle's works, thus having an impact on Islamic alchemy.

[139] Legend has it that Egyptian alchemy originated with the "Emerald Tablet," written by the Egyptian god of mathematics and science, Thoth (the Greeks give credit to the god Hermes, thus resulting in the combination Hermes-Thoth, or Thrice Great Hermes also known as Hermes Trismegistus). The Emerald Tablet is undoubtedly one of the oldest and longest-standing of all alchemical documents. It wasn't unusual for Greek and Islamic translators to credit someone (usually an earlier author) other than the original author for a given alchemical work. Indeed, the earliest surviving version of the Emerald Tablet appears in an early ninth-century Arabic text credited to Apollonios of Tyana, a first-century (AD) magician. The Emerald Tablet is written in cryptic form, as many alchemical writings were, and several have provided interpretations, including Newton. Alchemists in the Middle Ages and the Renaissance adopted a belief system following from the Emerald Tablet, sometimes referred to as the microcosm-macrocosm view, or hermetic tradition.

[140] Alexandria was founded in 331/332 BC by Alexander the Great.

experimental chemistry we know today. Thus alchemy was, in some sense, experimental chemistry's very humble beginnings.[141]

Most of us are familiar with the alchemist's goal of transforming the "lesser" base metals (mercury, lead, tin, copper, and iron) into the more precious metals of gold and silver (in a process known as chrysopoeia), a goal that persisted into the 1720s. While the idea of transmuting one material into another seems ridiculous to us today, it essentially followed from Aristotle's natural philosophy. Aristotle adopted the four-element theory of Empedocles, modifying it further by giving "qualities" to each of the elements: earth was cold and dry; fire was hot and dry; water was cold and wet; and air was hot and wet.

To the alchemist this meant that one merely needed to exchange one or both of the qualities of a given element to transform it into another element. For example, starting with water (cold and wet) and heating it would, to the alchemist, appear to transform it into air (hot and wet) as the water boiled and eventually evaporated away. Indeed, Aristotle did believe in such transformations among the elements. Moreover, he believed every substance to be composed of all the elements, with the difference between substances resulting only from the differences in the proportions of the elements within them.

---

[141] Since I was a chemistry major in college, I had to take several lab classes. For each of these labs I was required to keep a lab notebook to record all my observations of the experiments performed. A fundamental rule for keeping the lab notebook was to never tear out any of the pages if a mistake was made. Rather, simply crossing out the mistake with a single line was enough. In this way, I would be able to revisit mistakes later on if I thought there was still something useful to be found. Sometimes the mistake or the process of making it leads to something useful, and therefore forgetting it altogether isn't such a good idea. Indeed, this was often the case, and being able to take a closer look at my initial "mistake" often provided useful insight. In all, my final notebook would always contain my correct results along with my "mistakes." In some ways, one can think about the relationship between alchemy and chemistry in this way, not that I'm strictly categorizing alchemy as a blatant mistake. Rather, the similarity is that the initial effort that one thought to be not quite right at first can lead to the right answer (or some form thereof). Often the answer was actually present in the "mistake" upon further inspection or with a little more effort put in.

This idea was extended with the work of the Islamic alchemist Jabir ibn Hayyan (c. 721 AD–c. 815 AD).[142] Jabir followed Aristotle's lead but replaced his concept of qualities with "natures." However, his most significant work was his "mercury-sulphur" theory of metals. Jabir imagined metals to be composed of mercury and sulphur. Not just any old mercury and sulphur, mind you, but rather special forms that were only approximated by the ordinary mercury and sulphur we know today. Jabir felt metals differed only by the purity of the sulphur and mercury they contained and the ratio thereof. Therefore, gold was the manifestation of the purest form of sulphur and mercury in the perfect ratio.

However, since all metals contained sulphur and mercury, one should be able transform any one of them into gold by using a *catalyst*, or elixir, which Jabir called in Arabic *al-iksir*. Generations of alchemists inherited this theory, and in Western alchemy, this elixir became known as the "philosopher's stone," which fundamentally was endowed with the power of separating and rearranging the fundamental constituents of matter, thereby allowing the creation of something totally new.

In the Middle Ages the philosopher's stone became something more than just a tool for transforming the base metals into the precious metals of gold and silver. It became connected with an elixir of life that would restore the diseased body back to perfect health, thus providing longevity. Therefore, in general, the philosopher's stone became associated with the ability to transform something of "lesser quality" into something of "perfect quality": base metal into precious metal, or diseased body into healthy body.

For the medieval alchemist, this duality integrated well into the "microcosm-macrocosm" view (or hermetic tradition) in which man (the

---

[142] Jabir ibn Hayyan, whose Latinized name is Geber, reached a level never surpassed in Islamic alchemy. He is credited with originally making *aqua regia* (Latin for royal water), a mixture of nitric and hydrochloric acids (usually one part nitric and three parts hydrochloric by volume of the concentrated acids). *Aqua regia* dissolves a variety of metals, including gold and platinum, which are members of the group of metals known as the "royal metals" (hence "royal water"), or "noble metals," that are highly resistant to corrosion and oxidation. Nonetheless, *aqua regia* can't dissolve all the royal metals.

microcosm) was intimately connected to the universe (the macrocosm). Thus, the process of purifying a lesser metal into gold was believed to be the same technique needed for the purification of the soul.

Paracelsus (1493–1541)[143] reworked Jabir's mercury-sulphur theory of metals into a version he called the *tria prima*, where he believed all matter (not just metals) to be composed of the "spiritual principles": salt, sulphur, and mercury. These were as much symbolic categories (and had associated qualities just as the four elements did in Aristotle's theory) as they were the fundamental constituents of matter. Once again, Paracelsus was not talking about the salt, sulphur, and mercury that we recognize today.

Paracelsus was a pioneer in using alchemy for medicinal purposes, which he believed to be alchemy's main purpose: "Many have said of Alchemy, that it is for the making of gold and silver. For me such is not the aim, but to consider only what virtue and power may lie in medicines."

For Paracelsus, health and disease were related to the human body's connection to the universe. Therefore, Paracelsus viewed a medicine as something that restored the necessary "harmony" between the body and universe. The alchemist is often viewed as a thief, cheat, or scoundrel, but the fact of the matter is that alchemy was practiced by people of all occupations, and often attracted the patronage of kings.

The very educated and brilliant Isaac Newton began studying alchemy around 1669 and continued for some thirty years. His alchemical manuscripts total well over a million words (the Bible is 773,692 words), and 138 alchemy books were part of his personal library. Thus, although we know him for his works in establishing the area of physics referred to as classical mechanics or Newtonian mechanics (discussed in his *Principia*), he probably wrote more about and spent more time on alchemy.

The appeal of alchemy for Newton was that, unlike the mechanical philosophy that essentially eliminated God's role from the workings of the universe, alchemy almost demanded a spiritual presence of some sort as

---

[143] Philippus Aureolus Paracelsus, or simply Paracelsus, was born Theophrastus Bombastus von Hohenheim.

part of its philosophy. We get a sense of this from the microcosm–macrocosm views of the medieval alchemist and from Paracelsus' view of health and medicine. Moreover, Newton probably saw alchemy as a complementary and less restrictive view of nature than was provided by the mechanical philosophy.[144]

In the seventeenth century, the science we now call (experimental) chemistry was merely beginning, whereas alchemy was still practiced by many. However, historically it's very difficult, perhaps impossible, to clearly tell when alchemy ended and chemistry began. In fact, these words were pretty much interchangeable until around the end of the seventeenth century. It wasn't until the early eighteenth century that "alchemy" and "chemistry" had acquired their modern meanings, with "alchemy" referring exclusively to the process of attempting to change "lesser" metals into gold and silver.

Eventually, Aristotle's four-element theory would fade from the picture, as would Paracelsus' tria prima theory with its "spiritual principles" of salt, sulphur, and mercury. The ancient atomic theories of Democritus and Epicurus would finally gain favor. Alchemy would finally concede to chemistry as the experimental procedures it initiated gave way to rational experimental design. Moreover, the mechanical philosophy would provide a sort of backdrop for all of these things as a new generation of scientists came to believe that nature could indeed be understood rationally as working much like a machine with specific laws governing its every action: a world machine.

---

[144] It would be hard to imagine that alchemy didn't shape some of Newton's ideas, perhaps even some of the "more respectable" ones we know him for. Indeed, it seems that his studies in alchemy led him to appreciate something very fundamentally and ultimately important about atoms: they are affected by both the forces of attraction and repulsion; they "push" and "pull" at each other in varying degrees. There has even been speculation that his knowledge of the attractive nature of atoms led him to his law of gravity. We'll never know this for sure.

CHAPTER 11

Realizing Atoms

The Physical Foundations of the Atom

Through the eighteenth century, the words *particle, corpuscle, element,* and *atom* were all used synonymously to refer to the building blocks of matter. In fact, no more insight into what an atom was had been advanced since Democritus' description some two thousand years earlier. For the chemist, the atom as the fundamental building block of matter was conceptually nice but offered nothing towards understanding his main concern, which was how matter undergoes various reactions.

Consider for a moment that you're a chemist in the late eighteenth century who takes his starting materials (the reactants), proceeds to mix them together, and watches as the chemical reaction occurs and produces something new (the products). How does knowing that the atom may be the building block of matter provide you with anything useful in explaining the chemical reaction you just witnessed?

To many chemists of the time, the atom didn't appear to play a significant role in their daily experiments. Sure, chemists would draw "atomic pictures." However, for most chemists these were nothing more than a visualization tool, useful in organizing certain thoughts about a chemical reaction, but not in providing a detailed explanation of the actual process. What chemists really wanted was something to explain the vast amounts of accumulating experimental data.

How was knowledge of the atom (if it even existed) actually going to provide this kind of information? Moreover, if the atom was indeed the smallest, indivisible part of matter, how would you know if you had it? How much did it weigh? Are there different types of atoms? These were some of the questions on chemists' minds.

With all the different atomic theories floating around, the atom held very little distinction and offered more confusion than clarity. What was desperately needed was a clear path towards answering some of those questions. The first sense of such clarity began by introducing a more precise version of the atom concept called an *element*.

Elements and Atoms

In 1661, Robert Boyle wrote his most important work, *The Sceptical Chymist*. In it he brutally attacks the element theories of Aristotle and Paracelsus. More importantly, he proposed the idea of an *element* as something that can't be further broken down by any experimental means:[145]

> ... certain primitive and simple, or perfectly unmingled bodies; which not being made of any other bodies, or of one another, are the ingredients of which all those called perfectly mixt bodies are immediately compounded, and into which they are ultimately resolved.

---

[145] The timing couldn't have been better because a new trend in experimental investigations was beginning. The days of simply making observations and recording data was giving way to a more critical process of experimentation. In 1620, Francis Bacon (1561–1626) wrote *Novum Organum*, which demanded a new approach to gathering and interpreting experimental data. In it he argues for an *inductive* rather than *deductive* approach to dealing with experimental data, and thus provided the starting pointing for the scientific method that we use today. He also argued that to really gain fundamental insight into nature one needed to put nature on the spot, so to speak. Bacon says it quite elegantly: "under constraint and vexed; that is to say, when by art and the hand of man she is forced out of her natural state, and squeezed and moulded." Indeed, the Aristotelian approach of merely observing things in nature from afar was giving way to setting up experiments that allowed for the controlled observation of natural phenomena.

More than a hundred years later, in 1789, Antoine Lavoisier reiterates this in *An Elementary Treatise on Chemistry*:

> if we apply the term *elements* ... to express our idea of the last point which [chemical] analysis is capable of reaching, we must admit, as elements, all the substances into which we are capable, by any means, to reduce bodies by decomposition.

Of course, they were referring to the means available at the time. Thus, whether future generations of scientists would devise new methods to separate further those things that they had labeled as elements was something only time would tell. However, this was not the primary concern, nor should it have been. As Lavoisier goes on to say:

> Not that we are entitled to affirm, that these substances we consider as simple may not be compounded of two, or even of a greater number of [elements]; but, since these [elements] cannot be separated, or rather since we have not hitherto discovered the means of separating them, they act with regard to us as simple substances, and we ought never to suppose them compounded until experiment and observation has proved them to be so.

Lavoisier was aware that sometime in the future things that were then thought to be elements might actually turn out to be more complicated substances. These substances, being separated further by newer methods, would finally reveal the actual elements constituting them. However, the importance lies in the definition of *element* as stated by both Boyle and Lavoisier.

As already mentioned, the reality is that back then the concept of atom was vague. However, by establishing a physical identity for atoms, the concept of an element started to clear the way. An element is an atom of a certain type – it's really that simple.

The unique identity of an atom assigned to it by its element determines the way it will interact with other elements. Our knowledge of these interactions provides the basis for understanding chemical reactions in a meaningful way, and it's through these chemical reactions that matter is "built up" and "broken down."

As of December 2015, we know of 118 elements, whereas Lavoisier identified thirty-three in *An Elementary Treatise on Chemistry* published in 1789, which included caloric (heat) and light. He also included the modern-day elements: oxygen, nitrogen, and hydrogen. Other common examples of elements that we are familiar with today are gold, silver, iron, copper, tin, and mercury.

The Way Elements Combine

In 1789, Lavoisier showed that the total mass (or matter) is conserved – it doesn't change from beginning to end – during the course of a chemical reaction.[146] In other words, if one carefully weighs out one gram of starting materials (reactants), the amount of ending materials (products), after the chemical reaction has finished, will be one gram.[147] It's that simple. No mass gained, no mass lost during the course of a chemical reaction. So, what exactly happens during the course of a chemical reaction?

---

[146] Lavoisier was very meticulous when performing his experiments. He emphasized quantitative experimental methods and the use of the chemical balance to carefully weigh the starting material and ending material. His chemical experiments disproved the phlogiston theory, which stated that phlogiston was something that was stored inside a substance, and that the act of burning it (combustion) caused it to be released. However, Lavoisier's experiments showed that when a substance burns, it combines with oxygen in the process; burning or combustion is a process that involves your starting material combining with oxygen. Therefore, when the starting material combines with oxygen you get your ending material, which is actually heavier than your starting material. (In this example, we aren't counting oxygen as a starting material.) If indeed fire was a release of phlogiston, one would expect combustion to result in a decrease in mass, whereas Lavoisier had clearly shown it increased.

[147] However, sometimes there are starting materials left over, which is to say that not all of it was used up during the course of the chemical reaction. In this case, the amount left over along with the amount of ending materials have to equal the original amount of starting materials.

Apparently, "things" simply "reorganize" themselves. After all, if nothing is lost or gained, but what you end up with is different from what you started with, then all that is left is to shuffle things around – right?

Think of it like a deck of cards. Let's say we have a brand-new deck of cards right out of the package with all the cards in order (starting materials or reactants), and now we shuffle the deck of cards (chemical reaction), ending up with something new (ending materials or products). Clearly, this new thing is simply a reorganization of the original deck of cards. That is to say, no cards have been lost or added (unless I am cheating in a card game perhaps); but rather, it's simply that the order of the cards has changed.

So, a chemical reaction is sort of like a deck of cards being shuffled. As far as the things that reorganize, John Dalton (1766–1844) concluded these were the atoms of the elements making up the starting materials. So the next natural question really must be: If we want to perform a chemical reaction, how much starting materials is needed to get the desired ending materials?

In other words, is the actual amount important, or can any old amount of starting materials be tossed in to obtain the desired ending materials? Perhaps chemical reactions are like cooking, where you simply add a "bit" of this, or a "dash" of that. Claude Louis Berthollet (1748–1822), a disciple of Lavoisier, thought just this.

When baking an apple pie one might use five apples, six apples, and so on, but most of us would still simply call it an "apple pie" in the end, without describing it as a "five-apple apple pie" or a "six-apple apple pie." This was how Berthollet imagined elements to combine during the course of a chemical reaction to make a compound.

Berthollet was a respected scientist whose ideas carried weight at the turn of the nineteenth century. He believed that the elements making up a given compound didn't occur in a particular ratio to each other, but could occur in varying proportions and still result in the same compound – it was all "apple pie" to him. Moreover, according to Berthollet, the proportions of the elements in a compound were determined by the amount of starting materials used. Therefore, if the starting materials were such that some elements occurred in greater amounts than other elements, the resulting compound was expected to have more of these elements present in it as well. Like in cooking, if more sugar is added to a recipe, the final dish is expected to be sweeter.

This concept didn't go over well with Joseph Louis Proust (1754–1826), who didn't view chemistry as a culinary art. Proust had conducted detailed studies in a well-equipped laboratory and found that the elements forming a given compound are not variable but are in an exact and fixed proportion to each other. Thus, different proportions of these same elements resulted in different compounds, not simply variations of the same compound as Berthollet claimed – so it does matter whether you're using five apples or six apples, and it's not all apple pie when doing chemistry. Proust makes it pretty clear:

> the properties of true compounds are invariable as is the ratio of their constituents [elements]. Between pole and pole, they are found identical in these two respects; their appearance may vary owing to the manner of aggregation, but their properties never.

Proust put forth the *Law of Definite Proportions* in 1799 and then fought with Berthollet over it until 1808, when Berthollet finally conceded to it. It seems Berthollet might have been led to his conclusions, in part, by analyzing impure compounds and mixtures. The Law of Definite Proportions, or *Proust's Law*, has some far-reaching consequences, as will become clearer in a minute. However, one in particular to note is that whether you prepare a compound in the lab or it's found occurring in nature, the proportions of the elements that go into making that compound are the same in both cases; the method of preparation of a given compound does not change its chemical makeup.

This may seem trivial to us now, but this was not common knowledge in the early nineteenth century, and Proust's research and others have assured us of this important fact. The Law of Definite Proportions and *Lavoisier's Law of Conservation* formed the critical backdrop when Dalton began working on his atomic theory.

Modern Atomic Theory: The Beginning

John Dalton, the son of a hand-loom weaver, was born into a Quaker family at Eaglesfield in Cumberland, England. At the age of twelve, the Quaker school he was attending was turned over to his older brother, who called upon John to assist in teaching; two years later, the brothers bought a school of their own.

In 1793, Dalton moved to Manchester to teach mathematics at the New College. Dalton's original interests were in meteorology, and he published *Meteorological Observations and Essays* that same year. But Dalton had various scientific interests and soon turned to the study of gases, an interest he acquired upon pondering the Earth's atmosphere, which is a mixture of gases we call air.

Dalton wondered why the experimental data of the time indicated the Earth's atmosphere was *homogeneous*. In other words: Why did it show very little variation in composition with increasing altitude? He knew that the atmosphere of the Earth was a mixture of gases of varying densities. (Today, we know the Earth's atmosphere to comprise 78.09% nitrogen, 20.95% oxygen, 0.93% argon, 0.039% carbon dioxide, and small amounts of other gases. Air also contains a variable amount of water vapor depending on temperature, averaging around 1%.) Thus, Dalton expected the denser gases to persist at the lower altitudes, while the less dense ones dominated at higher altitudes. This makes some sense, really.

After all, oil floats on top of water because it's less dense than water. Moreover, a balloon filled with helium flies high to the top of the Earth's atmosphere because helium is less dense than air. So why shouldn't the less dense gases float on top of the denser ones in the Earth's atmosphere? Although unknown at Dalton's time, the reality is that the composition does actually vary in height. Nonetheless, it isn't very noticeable at only a few miles above the Earth (which was the extent of the experimental data in Dalton's time) but does become quite noticeable at higher altitudes. Anyone who has been at higher altitudes (perhaps snowboarding or hiking in the mountains) can attest to breathing a bit heavier as a result of the lower amounts of oxygen.

Oxygen happens to be one of the denser gases making up air, and therefore you do find less and less of it as you move farther and farther upward into the Earth's atmosphere. Nonetheless, a mixture of gases, like the atmosphere, is a bit different from a mixture of liquids, like oil and water. While mixtures of liquids of varying densities tend to separate themselves in layers such that the less dense ones are floating on top, gaseous mixtures are a bit different. The particles (either atoms or molecules) of a gas move farther and faster (in a given interval of time) than those of a liquid. As they do so, a single particle undergoes about a billion collisions every second with other particles. As a consequence, mixtures of gas tend to, well, mix. In fact, they tend to mix such that the particles are (more or less) evenly distributed throughout (or homogenously as we said before) without forming well-defined layers like liquid mixtures do.

Therefore, the Earth's atmosphere is a uniform mixture of gases whose density varies with altitude, getting less dense as you go up. Recall that we talked about the idea of particles in motion as the beginnings of kinetic theory. However, Dalton simply didn't believe in kinetic theory. He believed that atoms of an object remained fixed in place (a static model of matter). He also believed that the atoms of an object were always in direct contact with each other. This was consistent with Dalton's refusal to accept the concept of "action at a distance," believing instead that objects can only exert a force on each other if they are touching.

Dalton wasn't alone in his rejection of action at a distance.[148] Indeed, this was important in shaping Dalton's atomic theory. So how did Dalton

---

[148] In 1644, Descartes wrote his major work, *Principles of Philosophy*. Here, he makes clear his rejection of action at a distance, and he proposed that all of space is filled with pieces of matter that interact via – you guessed it – direct contact. In 1687, Newton wrote *Principia* and in it his theory of gravity was one of action at a distance where objects, such as the planets, feel gravity through empty space without anything mediating this interaction directly. Newton accepted action at a distance perhaps only half-heartedly. With Einstein's General Theory of Relativity, gravity is understood as resulting from the curvature of spacetime. Thus, matter curves spacetime with gravity arising from this curvature, spacetime affects the motion of matter, and action at a distance is eliminated. Action at a distance appears in *quantum entanglement*, where Einstein called it "spooky action at a distance," which we'll discuss when we talk about quantum mechanics in Part IV.

explain the mixing of gases in the atmosphere? Well, quite simply he concluded that the repulsive forces[149] between the particles of gas must have been responsible for mixing them, until at some point they came to equilibrium, remaining fixed in place and in direct contact ever since.

Dalton's interest in gases may have started with his inquiries into the Earth's atmosphere, but it soon became a more general interest into the behavior of gases.[150] Dalton became convinced that the solubility (ability to

---

[149] Dalton believed that particles of gas interacted through a force that caused them to repel each other and that this interaction increased as the particles got closer to one another. He adopted this belief from a short passage Newton wrote in *Principia* describing a calculation he had done showing that Boyle's Law could correctly be deduced from gas particles that interacted in this way. However, Newton himself didn't intend to imply that gases really do interact in this way.

[150] Dalton is also known for his law of partial pressures, which is a law for the total pressure exerted by a mixture of gases in a container. It's accurate when the gases in the mixture don't strongly interact with each other – that is, where the gases in the mixture exhibit *ideal gas* behavior. According to *Dalton's Law of Partial Pressures*, the total pressure of the mixture is simply found by adding up the partial pressures of each of the individual gases in the mixture. The partial pressure is the pressure a given gas of the mixture would exert if it were alone in the container; that is, each gas in the container is assumed to behave independently of the other gases of the mixture. Dalton's Law of Partial Pressures can be understood from the *kinetic theory of gases*. According to the kinetic theory of gases, the average *translational kinetic energy* (the kinetic energy associated with movement in x, y, and z directions) of an individual particle (atom or molecule) in a gas depends only on the temperature; if the temperature is known, the average translational kinetic energy of a single particle in the gas can be determined! Therefore, at a given temperature all the particles in a mixture of gas have the same average translational kinetic energy. To be sure, if I have a mixture containing two gases, A and B, the particles of A have the same average translational kinetic energy as the particles of B. Thus, if I begin with a container of particles A and to it I add particles B, the average total kinetic energy will be the sum given by all the A and B particles. Now, the pressure of a gas in a container is a measure of the average total kinetic energy the particles are imparting to the container walls when they collide with them. Since the assumption is that the particles behave ideally, adding atoms B to the container of atoms A will increase the average total kinetic energy as the sum, and the total pressure will be the sum of the partial pressures of the gases A and B. It's rather ironic that Dalton's Law of Partial

dissolve) of a gas in water was related to the weight of the atoms of the gas.[151] As a result, Dalton shifted his focus to the determination of *atomic weights*.

As you can imagine, it's a bit tricky to determine the weight of an atom. After all, it's not as if you can see an atom, let alone simply set one on a scale and weigh it. The Law of Definite Proportions, as Dalton would have known it, insisted that when things come together to form a substance, or compound, they do so in a specific ratio to each other.

To be sure, the Law of Definite Proportions doesn't confirm the existence of atoms, but for Dalton it was a powerful statement in their favor. Dalton must have wondered why a compound can only be formed in such a way that the things constituting it only occurred in specific ratios. He must have been asking himself: What are these things, and why are the ratios fixed?

In the end, Dalton took a big step forward and concluded these things were the atoms (of the elements) making up the compound. There should be no mistaking it: Dalton considered atoms to be the physical constituents of matter, and an element (as we talked about before) gave an atom its identity and corresponding physical properties. This was a bold position to take at the beginning of the nineteenth century, when most didn't know

---

Pressures is understood from the kinetic theory of gases, a theory Dalton himself never believed in.

[151] Solubility is a measure of how well one substance (the solute) dissolves in another substance (the solvent). Consider a substance dissolved in water. Its ability to dissolve well in water is related to its ability to make "good" interactions with the water molecules. If the substance is unable to make enough good interactions with the water, it won't dissolve well. Common table salt dissolves well in water because its atoms make enough good interactions with water, whereas olive oil does not dissolve well because its atoms are unable to make enough good interactions with water. Therefore, strictly speaking, solubility is not necessarily related to the weight. However, there are some examples when a substance's weight does trend well with its ability to dissolve in a given solvent. Compounds made in drug discovery research (for potentially marketable drugs) often become less soluble in water as their weight increases. This occurs because as their weight goes up, so does the number of carbon atoms and their associated hydrogens. Quite simply put, this motif of carbon and its hydrogens does not make enough favorable interactions with water, thus lowering the solubility of the molecule as a whole.

what to make of atoms. With these starting points, he set out to determine the weights of the atomic elements.

Obtaining a system of atomic weights was a tough business in 1800 when Dalton began his task, and he needed to make a few reasonable assumptions, or hypotheses. First, Dalton not only decided that the Law of Definite Proportions implied that atoms combine in a specific ratio, but that they also do so only in whole-number ratios rather than fractional ratios. So the ratios would be, for example, two to one (2:1); three to four (3:4); etc. – not one to one-fourth (1:1/4); one-half to one-fifth (1/2:1/5); and so on. The reason for this is simple: Dalton believed atoms to be *indivisible*:

> Matter, though divisible in an extreme degree, is nevertheless not infinitely divisible. That is, there must be some point beyond which we cannot go in the division of matter. The existence of these ultimate particles of matter can scarcely be doubted, though they are probably much too small ever to be exhibited by microscopic improvements.

Therefore, believing in atoms as indivisible means you can't divide them in half, thirds, or fourths, etc., which means they must combine in simple whole-number ratios. It's that simple.

Recall that Lavoisier said that mass (or matter) is conserved over the course of a chemical reaction. Moreover, Dalton's interpretation of the Law of Definite Proportions was that during a chemical reaction, atoms combine in specific whole-number (not fractional) ratios to come together to form a compound. Therefore, with these two concepts in place, we further conclude that during a chemical reaction it's the individual atoms themselves that are being conserved; the atom is the *discrete* unit of mass conservation of a chemical reaction. Dalton's other main conclusions on the nature of atoms are as follows.

- **All atoms of a given chemical element are the same**. The concept of chemical elements as put forth by Boyle and Lavoisier and continued in Dalton's atomic theory was a main feature distinguishing it from the ancient Greek atomic theories that we discussed earlier. While atoms make up matter, not all atoms are the same. Rather, atoms differ from each other only with respect to the element they make up – the element "types" the atom, so to speak. Dalton wrote, "the ultimate particles of all homogeneous bodies are perfectly alike in weight, figure, etc."

Some of the elements already familiar to you (along with their chemical symbols) are perhaps oxygen (O), hydrogen (H), copper (Cu), lead (Pb), gold (Au), silver (Ag), and aluminum (Al), to name a few.

- **Atoms are unchangeable**. Centuries of failed attempts by the alchemists to change lead into gold undoubtedly convinced Dalton that it was impossible to change an atom of one element into that of another element. Indeed, an atom of a given element can't be changed into another element. For example, you can't change oxygen into hydrogen. However, these days we know that atoms are not so unchangeable (or eternal) – in fact, they "fall apart" by undergoing radioactive decay, which we'll discuss later. This feature of certain atoms was not known in Dalton's time, which was probably a good thing as it would have most likely further confused the issue of the nature of atoms.

- **Atoms combine to form bigger things, known as molecules**. Atoms are the fundamental building blocks of matter. However, a group of atoms can come together to form something just a bit bigger – not big enough to be seen with the naked eye, but still bigger than a single atom. These combinations of atoms form molecules or compounds.

- **In chemical reactions atoms only rearrange themselves**. We now know this follows from atoms being conserved during the course of a chemical reaction. Since atoms are not created or destroyed, they must simply rearrange ("shuffle") themselves to form – what else – molecules. Moreover (as just noted), they don't change their identity to become a different element in a chemical reaction, either.

These ideas formed Dalton's belief system on atoms and allowed him to develop a very impressive atomic theory of his own, which included a set of atomic weights for different elements. Let's take a closer look at how he was able to accomplish such a feat.

Atomic Weights of the Elements

Atoms are small, with weights ranging from 10^{-22} to 10^{-24} grams. Now consider an atom in the middle of that range, weighing 10^{-23} grams, which is 0.00000000000000000000001 grams. Now consider a grain of sand measuring 1/400th of an inch, which would weigh approximately 0.001 (10^{-3}) grams.[152] Therefore, a grain of sand is 100,000,000,000,000,000,000 times bigger than a single atom weighing 10^{-23} grams – no wonder you can't actually see an atom. So how did Dalton determine the atomic weights of some key elements such as oxygen, hydrogen, and nitrogen?

It should be clear that when one weighs any object, its total weight is related to the weights of all of the atoms it comprises. Clearly, the individual weights of all the atoms that make up the object will, without a doubt, total up to the weight of the overall object. If the object is made of only one type of atom (element), then it's even simpler: the number of atoms in the object multiplied by the weight of the element equals the total weight of the object.

For experimental purposes, a good starting point for an "object" made up of one type of atom would be a pure gas, like oxygen, nitrogen, hydrogen, etc. Moreover, in some cases we can take two different pure gases and mix them together such that they undergo a chemical reaction to form a product. For example, if hydrogen gas is mixed with oxygen gas, they will form water vapor (under the right conditions).[153]

From the conservation of mass/atoms, one finds, upon weighing the reactants (hydrogen and oxygen) and the product (water vapor), that the total amount of the reactants used and the resulting product formed will equal each other. The question, though, is how does any of this help one determine the weight of the atoms involved in this chemical reaction (namely oxygen and hydrogen)?

Let's say we determine the total amounts of oxygen and hydrogen used in the chemical reaction, which Dalton and others would have done using the scales available in the nineteenth century. The total amount of oxygen

---

[152] "Sand", World Book Encyclopedia (Chicago: World Book, 2000).

[153] This is a combustion reaction, which means you need to add a flame source to cause hydrogen to burn, thus triggering the reaction.

(or hydrogen) used in the chemical reaction is simply the number of oxygen (hydrogen) atoms used multiplied by the actual weight of a single atom of oxygen (hydrogen). That's it. Further, we can divide (for example) the total amounts of oxygen and hydrogen used to obtain the *ratio* of the number of atoms (of oxygen and hydrogen) used multiplied by the ratio of the actual weights of their single atoms.

This last ratio is essentially a *relative weight*, which we are very close to determining, although we still haven't successfully determined the actual weight of either an atom of oxygen or an atom of hydrogen. To be honest, the actual weights of the atoms are not really important if we can get their relative weight, which is the ratio between the actual weight of a given atom and some arbitrary *reference* atom's actual weight.

For example, our modern convention assigns a single atom of oxygen a relative weight of about 16, whereas a single atom of hydrogen has a relative weight of about 1. What this means is that an atom of oxygen is sixteen times heavier than an atom of hydrogen. However, this doesn't mean that oxygen's actual weight is 16 in some unit. Additionally, the reference atom for our current system is carbon, which has been assigned a reference weight of about 12. This means a carbon atom is twelve times heavier than a hydrogen atom, but it's only 0.75 times as heavy as an oxygen atom ($12 \div 16 = 0.75$).

I did say we were "close" to determining a relative weight, but we are not actually there yet. While we know the ratio of the total amounts of oxygen and hydrogen used in the chemical reaction (simply by weighing), we don't know the ratio **between** the atoms of oxygen and of hydrogen that combine to form water. If we did, we could then determine the relative weight between oxygen and hydrogen.

While it's common knowledge today that water is H_2O, Dalton had no idea. An incomplete amount of reliable experimental data and a poor understanding of how atoms come together to form products essentially prevented this from happening. This was the type of problem Dalton had to overcome, and he did so by implementing a very simple rule: *Dalton's Rule of Simplicity*. Dalton's rule was nothing more than an assumption, one that he needed if he was going to make any sort of progress in determining relative weights of atoms.

Dalton assumed that if two elements come together (such as oxygen (O) and hydrogen (H)) and they can only form one product, then their atoms combine in the simplest manner possible, which would be 1:1. Now, if these same elements come together to form two products, the rule becomes 1:1 for the first product, and 1:2 for the second product. This concept was extended even further so that the same elements forming a third product would then force one to conclude that their atomic ratio was 1:3. Once again, this was an assumption that sometimes worked and sometimes failed.

Recall our example with oxygen and hydrogen forming water vapor. In Dalton's time, water was the only product, or compound (or compound atom as Dalton called it), of the elements oxygen and hydrogen. Therefore, one product means a ratio of 1:1, making water HO, which is exactly what Dalton concluded. Today, we know this isn't correct: water is H_2O.

However, there were some cases where Dalton's rule did work. Consider carbon monoxide and carbon dioxide, which were the only two compounds known in Dalton's time formed by the elements carbon (C) and oxygen (O). Knowing that twice as much oxygen is used to form carbon dioxide than carbon monoxide, applying Dalton's Rule of Simplicity makes their respective formulas CO_2 and CO, which is correct. Notice that the ratio of oxygen between the two oxides of carbon is 2:1, while that of carbon is fixed, or simply 1:1.

This is an example of what Dalton called the *Law of Multiple Proportions*, which, simply stated, says that when the same elements come together to form multiple compounds, they do so in a manner that the respective elements are whole-number ratios of each other. In our example with the oxides, as already noted, carbon is in a ratio of 1:1 between the two compounds, while oxygen is in a ratio of 2:1, which are both simple whole-number ratios. By now, this shouldn't be a surprise to you. After all, this is simply another consequence of believing in atoms as being indivisible entities that come together to form compounds; if you can't divide atoms then you can't have a single compound with a fractional ratio between its elements, nor can you have multiple compounds formed by the same elements with fractional ratios among their respective elements. However, just as Dalton's Rule of Simplicity was too simple, so was his Law of Multiple Proportions.[154]

With a firm belief in atoms, impressive physical insight, and armed with a few simple rules, Dalton was able to construct a table of relative weights, which he first presented in 1803 at a talk to the Literary and Philosophical Society of Manchester. In 1805, this effort first appeared in print, with a systematic explanation of the method appearing in 1808 when Dalton published the first volume of his book *A New System of Chemical Philosophy*. Here, with hydrogen as his reference, he gave the following relative weights: hydrogen (H) 1; nitrogen (N) 5; carbon (C) 5.4; oxygen (O) 7; phosphorus (P) 9; sulfur (S) 13; and so on, including several elements and compounds.

Recall that, as relative weights, these only represent the weight ratios, rather than the actual weight of the element. For example, according to Dalton's table, nitrogen was five times heavier than hydrogen. A quick look at a modern periodic table of the elements reveals that Dalton got all the relative weights wrong (except for hydrogen). How did Dalton make so many mistakes?

Well, recall that he had no way of knowing the exact ratios atoms were combining in to form a given molecule, and he addressed this by implementing his (very arbitrary) Rule of Simplicity. Moreover, although Dalton had remarkable physical intuition, he was a pretty poor experimentalist, and thus the quality of his own data suffered. But what probably hurt Dalton the most was the fact that the substances he was working with – gases such as oxygen, nitrogen, hydrogen, and the like – are not *monoatomic* elements.

Dalton's strict belief in atoms as indivisible meant he viewed the gases he was working with as consisting of single atoms, or monoatomic in nature. In general, this is a pretty harmless assumption since the majority of elements do exist this way. Unfortunately, the gases Dalton was working with don't. Rather, they exist as two atoms "connected" together, or as *diatomic* molecules.

---

[154] Compounds containing several elements may fail to adhere to the Law of Multiple Proportions. For example, consider the organic compounds sucrose (table sugar) and glucose (blood sugar), whose molecular formulas are $C_{12}H_{22}O_{11}$ and $C_6H_{12}O_6$, respectively. Here, a simple whole-number ratio only occurs for carbon as 2:1.

For example, oxygen, hydrogen, and nitrogen exist in nature as the molecules O_2, H_2, and N_2, instead of the single atoms Dalton imagined. As more experimental data became available, it became clear that Dalton's results were wrong. Dalton himself tried to amend his atomic theory, and in the specific case of water he stated in 1810: "After all, it must be allowed to be possible that water may be a ternary compound" – that is, H_2O or HO_2 instead of his original speculation of HO.

At the time, there was simply no way of knowing the exact ratio the elements were combining in when forming a compound. The failure of Dalton's system of atomic weights shouldn't overshadow his initial tenets, which were pioneering for his time and still hold true today (well, mostly, as we'll discuss in more detail later).

One could imagine that by an iterative processing of all the experimental data perhaps Dalton could have brought agreement between his atomic weights and experiment by systematically eliminating inconsistencies, a daunting task for sure. Fortunately, new insight was just around the corner.

CHAPTER 12

Final Doubts to Rest

The Atom as Physical Reality

By the middle of the nineteenth century, the establishment of the first law and all the work leading up to it had dealt the final blow to caloric theory and unified the concept of energy. While not everyone was convinced, the popular belief that heat was due to the motion of the fundamental particles of matter cast atoms into an even brighter spotlight than before.

As the nineteenth century pushed forward, pioneering scientists such as Clausius, Maxwell, and Boltzmann built theories on the presumption of the physical existence of atoms. Then in 1905, a young Albert Einstein wrote a paper (while working as a patent clerk) on a well-known physical phenomenon of the time, *Brownian motion*, which was first noted in 1827.

Einstein's theory correctly described Brownian motion, and at the heart of his theory was the existence of atoms that were in constant motion. Moreover, Einstein's theory established the first experimental means for actually proving the existence of atoms. The experiments performed soon after eventually confirmed the predictions made by Einstein's new theory. That the atom as a physical reality of nature, rather than merely a convenient visualization tool, was now forever entrenched in modern chemistry and physics.

Combining Volumes

In 1808, Joseph Gay-Lussac (1778–1850) was reproducing the well-known experiment of getting water vapor from hydrogen and oxygen gases. He did this by mixing specific volumes of each of the gases and combusting the mixtures with an electric spark. What he noticed was that the volumes of oxygen and hydrogen used in the reaction (the combining volumes) occurred in a simple whole-number ratio. In fact, upon careful review of not only his own work but also that of others, he concluded that the volumes of gases involved in a chemical reaction are always in a simple whole-number ratio:

> It appears to me that gases always combine in the simplest proportions when they act on one another; and we have seen in reality in all the preceding examples that the ratio of combinations is 1 to 1, 1 to 2, 1 to 3.

Moreover, Gay-Lussac noticed that if the resulting product was also a gas at the temperature and pressure of the experiment, its volume was also in a simple whole-number ratio to the participating gases. For example, in the case of mixing volumes of hydrogen and oxygen to obtain water vapor he found:

> 2 volumes hydrogen gas + 1 volume oxygen gas → 2 volumes of water vapor

Gay-Lussac's Law of Combining Volumes has a rather familiar tone. After all, Dalton talked exactly this way, not about volumes, but of atoms involved in chemical reactions: atoms combine in simple whole-number ratios to form molecules during a chemical reaction. Thus, one might imagine that Dalton would have been excited to hear about these results, perhaps even finding them useful in reconciling the conflicts of his own atomic theory. But nothing could have been further from the truth.

Dalton had, very reasonably, assumed that different atoms were different sizes. He also believed that atoms in a gas were packed in tightly such that they directly touched each other. Since there was little to no space in between, the atoms were forced to remain motionless. Therefore, according to Dalton's model, if one wants to fill a balloon to a given size (volume), it will take more small atoms than large atoms to do so.

As an analogy, imagine an ordinary box (the balloon) that we want to place balls (the atoms) into until it's full. First, we use golf balls, packing them as tightly as possible. After noting how many golf balls it took to completely fill the box, we empty the box and begin again. However, this time instead of golf balls, we use basketballs. Once again, as with the golf balls, we pack the basketballs in as tightly as possible. As you would expect, it takes fewer basketballs than golf balls to completely fill the exact same box.

What troubled Dalton about Gay-Lussac's results was that they seemed to be saying the same number of atoms, regardless of the type of atom (the element) or its size, would always fill the same volume – the same number of golf balls or basketballs will completely fill the box. Worse yet, Dalton thought: What if Gay-Lussac's results were saying there's no such thing as small or large atoms, but rather atoms come in one size only?

Dalton refused to believe Gay-Lussac's results, taking consolation in his own experiments, which showed Gay-Lussac's results to be flawed. The reality was that it was Dalton's results that were flawed; these were delicate experiments and he was a clumsy experimentalist. However, a new theory seemed to offer some insight.

Avogadro's Number

In 1811, Amedeo Avogadro (1776–1856) (born Lorenzo Romano Amedeo Carlo Avogadro di Quaregna e di Cerreto) looked at Gay-Lussac's results and concluded that, at the same temperature and pressure, equal volumes of gas (like two balloons of the same size) contain the same number of "particles." These particles can be individual atoms, molecules, or even a mixture thereof.

So, if two balloons are filled to the same size, say one filled with helium and the other is simply blown up (and therefore is a mixture of oxygen, carbon dioxide, nitrogen, and water vapor), the number of particles in each will be exactly the same, since both balloons are at the same temperature and pressure (room temperature and atmospheric pressure). Avogadro wasn't the first to suggest this, but he was the first to formulate it into a very coherent and persuasive concept. *Avogadro's Hypothesis* has some very interesting implications.

If, at constant temperature and pressure, the same number of particles is contained in a given volume, then the theories of Dalton and Avogadro are in disagreement. Where's the problem? Recall that Dalton thought different types of atoms come in different sizes. This was a reasonable assumption, which today we know to be true. However, Dalton also imagined the atoms of the gas to be directly touching each other, which followed from his refusal to believe in action at a distance. Therefore, for Dalton, a balloon filled with gas meant that the entire space inside was completely filled with atoms packed in tightly against each other. Herein lies the problem. Let's consider our analogy again.

Earlier, we were concerned with packing in the balls (either golf balls or basketballs) completely until the box was full. This time let's see what happens if we don't worry about this requirement. However, instead of one box, we imagine two of the same kind. First, we place a basketball in one and a golf ball in the other. We continue to do this until at some point we can't fit anymore of either ball into its respective box. The box containing the basketballs will fill up first (since basketballs are bigger than golf balls and the boxes are the same size), therefore causing us to stop as promised.

In this way, we have ended up with two boxes of equal size (volume) that contain the same number of "particles." Of course, atoms do behave differently from basketballs and golf balls, but in this case (I assure you) our analogy is true to form. Therefore, if we could count the number of small and large particles placed into the two different balloons, we would find that at the point the balloons were the same size, not only would there be the same number of small particles as there are large particles in each balloon, but there's also a lot of empty space left over.

You see, the particles (atoms, molecules, or mixtures thereof) of a gas are not packed in tightly against one another. Rather, they are in motion, moving about in the empty space that surrounds them. Indeed, a model consisting of atoms in empty space is already familiar to us from the ancient Greek atomic theories of Democritus and Epicurus. Avogadro's Hypothesis of equal particles for equal volumes at constant temperature and pressure provided remarkable insight into the nature of gases. Avogadro never proved his theory, nor was he able to determine what the actual number of particles would be at a given temperature or pressure. However, Avogadro's Hypothesis follows from the kinetic theory of gases.

From the kinetic theory of gases, it can be shown that the average total *translational kinetic energy (KE)* of an ideal gas at temperature T, containing N particles, is given by the equation,

$$KE = \frac{3}{2}NkT$$

where k is just a fixed number known as Boltzmann's constant. The translational kinetic energy is the kinetic energy associated with movement in the x, y, or z (length, width, height) directions. This result, which can be directly calculated using the Maxwell distribution (as discussed in Part II), reveals the very interesting fact that the KE does not depend on the mass of the particles or their identity; rather, it depends only on the temperature. This means that at the same temperature, if two volumes of different gases contain the same number of particles, both volumes will have the same KE.

Another result from the kinetic theory of gases is that the pressure P of an ideal gas at volume V is given by the equation

$$P = \frac{2}{3}V \times KE$$

which is simply

$$P = NkT / V$$

after substituting our above result. Consider two different gases denoted as 1 and 2. Their respective equations for the pressure will be

$$P_1 = N_1 kT_1 / V_1$$

and

$$P_2 = N_2 kT_2 / V_2$$

Therefore, if two different ideal gases are at equal volume, pressure, and temperature, they will have the same number of particles since we then have $P_1 = P_2$, where both the volumes and the temperatures are the same, giving

$$N_1 kT / V = N_2 kT / V$$

or simply

$$N_1 = N_2$$

which is Avogadro's Hypothesis. However, real gases differ from ideal gases in that they do experience attractive and repulsive interactions. As a consequence, real gases only contain Avogadro's number of particles when they behave "ideally," which is at low pressures and/or high temperatures.

Today, we recognize Avogadro's Hypothesis with a fundamental constant called *Avogadro's number*, which is the number of particles in one *mole* (often written as mol) of a substance. One mole of any substance is simply the amount (in grams) of the substance that is equal to its atomic weight. For example, the atomic weight of carbon is 12 grams per mole, and therefore one mole would be equal to 12 grams. Another example would be water, which has an atomic weight of 18 grams per mole, and therefore 18 grams of water (or about one tablespoon) would be one mole. Whereas Avogadro was simply talking about the number of particles in a volume of gas,[155] Avogadro's number refers to the number of particles in a specific amount of substance (one mole) and applies to gas, liquid, and solid.

---

[155] In 1865, Johann Loschmidt (1821–1895), using kinetic theory, estimated the size of a single molecule of air. A simple manipulation of his equations would have allowed him to estimate the number of air molecules in a given volume. However, he didn't take this final step, which was later done by James Clerk Maxwell (1831–1879), who found a value of 1.9×10^{25} air molecules per cubic meter.

In 1909, Jean Baptiste Perrin (1870–1942) provided the first accurate experimental determination of Avogadro's number, from studying Brownian motion (we'll discuss this in more detail later), to be 6.7×10^{23} particles/mol. He was the first to relate it to the mole amount of a substance and suggested naming it after Avogadro. Today, the value is more accurately determined at about 6.022×10^{23} particles/mol.[156] An ideal gas at 32°F and atmospheric pressure would fill up a volume (like a balloon) to 22.4 liters, and would contain exactly 6.022×10^{23} particles.

Avogadro thought that a gas consisting of a single element (like oxygen, hydrogen, nitrogen, etc.) could consist of molecules, whereas Dalton insisted that they could only exist as single atoms. A common belief at the time was that atoms of like elements would repel each other, whereas those of different elements would attract each other. Therefore, Avogadro's ideas were in direct conflict with this line of thinking.

Further, if atoms of like elements do actually attract each other, it was not clear what would prevent them from simply condensing (coming together) to form a liquid. Today, we understand that there are atomic interactions of varying types and strength, making it possible to have like elements attract one another in specific numbers.

Avogadro's Hypothesis and the fact that like elements in a gas can form molecules allow us to explain Gay-Lussac's Law of Combining Volumes. Recall Gay-Lussac's result for the formation of water vapor from the gases oxygen and hydrogen:

2 volumes hydrogen gas + 1 volume oxygen gas → 2 volumes of water vapor

Recall that Dalton thought the correct equation for the way the individual atoms in the volume combined was:

1 atom of hydrogen + 1 atom of oxygen → 1 molecule of water vapor

---

[156] Since 2006, the U.S. National Institute of Standards and Technology (NIST) has recommended $6.02214179(30) \times 10^{23}$ particles/mol.

In other words, Dalton thought the formation of water vapor occurred by a single atom of hydrogen combining with a single atom of oxygen to form a single molecule of water. Therefore, if we have a volume containing several atoms of oxygen and several atoms of hydrogen, the expectation, from Dalton's statement, is that this process will occur until we've run out of atoms to pair up.

Let's attempt to bring agreement between Dalton's equation and Gay-Lussac results starting with the same number of atoms in each equal volume of gas (at the same temperature and pressure), as prescribed by Avogadro's Hypothesis.

In Figure 12.1, we start with the correct number of volumes for each gas involved in the chemical reaction as determined by Gay-Lussac, while simultaneously imposing Dalton's equation for the way the individual atoms actually combine. Further, since we are using Avogadro's Hypothesis, the number of atoms in each volume must be the same, but the actual total number of atoms in each is not important.

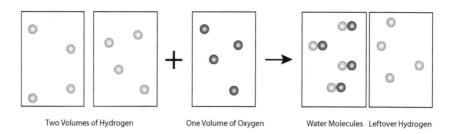

Two Volumes of Hydrogen   One Volume of Oxygen   Water Molecules   Leftover Hydrogen

Figure 12.1. Following Gay-Lussac's findings, we have two volumes of hydrogen combining with one volume of oxygen to make one volume of water vapor. Further, we implement Dalton's equation for the way he imagined the atoms in the volumes combined. Since the volumes, pressure, and temperature are taken to be the same, we apply Avogadro's Hypothesis by keeping the same number "particles," which in this case are single atoms in each volume.

Apparently there's a problem, since our simple procedure has resulted in leftover atoms of hydrogen. This would be fine if we knew this actually happened, but the experimental evidence tells us it doesn't. Let's remove

Dalton's equation from our procedure – after all, this was just speculation on his part. Instead, we allow for the possibility that atoms of like elements in the gaseous state can come together to form molecules as Avogadro suggested, rather than only existing as single atoms as Dalton demanded.

Specifically, let's assume that in hydrogen and oxygen two single atoms come together to form the *diatomic* molecules H_2 and O_2 respectively, rather than simply existing as the individual atoms H and O. Now we proceed as before, using only Gay-Lussac's results for combining volumes and Avogadro's Hypothesis (Figure 12.2).

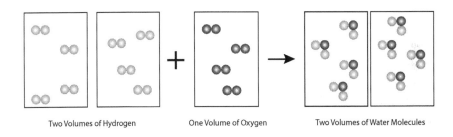

Two Volumes of Hydrogen   One Volume of Oxygen   Two Volumes of Water Molecules

Figure 12.2. Once again we consider Gay-Lussac's results in tandem with Avogadro's Hypothesis. However, this time we ignore Dalton's equation for the atom combinations and instead assume that hydrogen and oxygen exist as the *diatomic* molecules, H_2 and O_2, respectively. This allows us to satisfy Gay-Lussac's results and Avogadro's Hypothesis, and leaves us with no "leftover" atoms as before.

Since we don't have Dalton's equation for the atom combinations, we simply fall back on Avogadro's Hypothesis and enforce the same number of particles in every volume. Our only other requirement comes from Gay-Lussac's results that remind us that all the atoms of hydrogen and oxygen we start with will be used to form the final product of water vapor; in other words, we can't have any leftover atoms as before. In this way, we find something very interesting. If indeed we believe Gay-Lussac's results and Avogadro's Hypothesis, we see that the water molecules must have the formula H_2O – not HO as Dalton had speculated.

However, we kind of cheated; after all, we assumed that oxygen and hydrogen exist as diatomic molecules. This is something we know to be true today but was not known in Dalton's time. Even with eliminating Dalton's assumption and implementing Gay-Lussac's results along with Avogadro's Hypothesis, we are still left with a variety of possibilities. All we did was find one that works quite well (and is known to be right today, of course). So the question remains: How do we reconcile any of these methods and results?

While Gay-Lussac's Law of Combining Volumes and Avogadro's Hypothesis didn't uniquely solve our problem of how hydrogen and oxygen combine to form water, they did force us to reject Dalton's assumption of how atoms combine (Dalton's Rule of Simplicity). Moreover, one finds that by studying several reactions, rather than only one as we did, agreement between the various reactions can be reached, resulting in the correct chemical equations. In fact, Avogadro had a simple rule of his own. He said the correct equation for a given chemical reaction would be consistent with the equations of other related chemical reactions, Gay-Lussac's results, and his own hypothesis.

Avogadro's Hypothesis was a whole new way of viewing the atoms in a gas. That gases at equal volumes, temperature, and pressure would have the same number of particles, whether atoms or molecules, also meant that there must be a significant amount of empty space in the gas, and that the atoms weren't in constant contact with each other. Further, by proposing that the atoms of a gas can actually come together to form molecules challenged the popular idea that like atoms can only repel each other.

In 1811, these ideas weren't too mainstream, and Avogadro didn't offer reasonable explanations, either. He didn't calculate or experimentally determine the number of particles occupying a given volume at constant pressure and temperature to show that it was the same regardless of the particles. Consequently, Avogadro's ideas remained neglected for almost half a century.

The concept of the atom and the role it actually played in chemistry was still an open debate. In general, it was agreed that conceiving of matter as being made up of atoms provided a useful tool in chemical reactions and in visualizing the structure of the molecules they formed. As to whether this meant the true nature of matter consisted of ultimate, indivisible particles that should be denoted as atoms was another story. The lack of an unambiguous method for determining relative (and, of course, absolute) weights of atoms and molecules and their chemical formulas resulted in several incompatible atomic theories. Nonetheless, atomic theory maintained its foothold in chemistry in one form or another.

A much-needed turning point came in 1858 (two years after Avogadro's death) when Stanislao Cannizzaro published a pamphlet showing that Avogadro's work, aside from some minor exceptions to more general rules, could provide the basis for the determination of relative weights for many substances existing in the gaseous state. Recall that Dalton's approach required knowing the amount of starting materials used in a chemical reaction to form the molecule of interest and a guess as to how many atoms made up the molecule. The approach proposed by Cannizzaro reduced finding relative atomic weights to the almost trivial measurement of determining relative gas densities.[157] Unfortunately, Cannizzaro's pamphlet excited very few in the scientific community. This would soon change.

In 1860, Cannizzaro spoke at an international chemical conference held in the German town of Karlsruhe. His talk made a lasting impression on the audience, consisting mostly of prominent European chemists. Further, Cannizzaro's friend Angelo Pavesi distributed Cannizzaro's pamphlet to the attendees. Cannizzaro's system, based on Avogadro's Hypothesis, was adopted soon afterwards.

Cannizzaro's success in establishing Avogadro's work as paramount in atomic theory was due in part to his much-clearer account. However, perhaps the biggest factor was that, unlike Avogadro, he provided an (almost trivial) experimental means to test the hypothesis. Therefore, what was once mere speculation could now be readily verified and implemented. Further, the timing couldn't have been better.

Atomic theory had changed since its early days. By now the first law had been established (as of around 1850), which put an end to heat being

---

[157] This also means that in order to find the relative weight of the element or compound of interest, it has to either exist in the gaseous state or be amenable to being turned into a gas; this isn't possible for all substances, and therefore other methods have to be applied.

viewed as a fluid of particles (known as caloric); no more were atoms imagined as being surrounded by a layer of caloric (as Dalton proposed). Kinetic theory was coming into its own with the works of Clausius, Maxwell, and later Boltzmann.

Maxwell's revolutionary publication of 1860 described atoms of gas as moving with velocities that occurred within a well-defined range, or distribution. Thus, not only are the atoms in a gas not fixed in place as Dalton envisioned, they also move at different velocities. Now, with a consistent system of relative weights, it became possible to construct the periodic table of the elements, which organized the elements into groups or families, thus revealing certain trends in their properties. An atomic theory that finally worked with the experimental data provided chemists with a means of explaining chemical reactions, and the ability to write chemical formulas seemed to have finally arrived.

Brownian Motion

The initial work of Dalton and others after him seemed to establish a secure place for the atom as the indivisible particle constituting all of matter. By the 1860s, a chemical theory with the atom at its centerpiece now existed and, much to the delight of chemists, it provided the means to understand chemical reactions in a quantitative way.

The new atomic theory allowed for the accurate determination of relative weights. This was always the primary goal for a successful atomic theory, but as hard as Dalton tried, his method required a considerable amount of guesswork. By reviving Avogadro's Hypothesis, Cannizzaro was able to finally establish a reliable system for determining relative weights for many atoms and molecules. This was a game changer since, among other things, having reliable relative weights meant that one could determine the exact number of atoms that combined during a chemical reaction to form the final product. Indeed, a powerful atomic theory (for chemistry) was beginning to emerge.

While it was hard to deny the success of the atom concept in chemistry, there wasn't any real way of knowing if it wasn't all just mathematical fiction.

That is to say, while the idea of atoms provided a powerful means of writing and balancing chemical equations and rationalizing molecular formulas, there still wasn't any means of experimentally determining the presence or absence of atoms. This lack of experimental verification left many unwilling to accept the atom as physical reality although oftentimes willing to accept it as a useful tool. The atom's position as the ultimate particle of matter was still in jeopardy and would remain so until the work of two physicists, one a theoretician and the other an experimentalist, changed everyone's minds, once and for all.

In 1900, Albert Einstein (1879–1955) graduated from Zürich Polytechnic Institute (known as Swiss Technical University or ETH since 1911). An anticipated assistantship there never materialized, and it left Einstein job hunting. After two temporary positions, in 1902 (with the help of a friend's father) he landed a job as a technical expert third class at the Swiss Patent Office in Bern. For Einstein, the work at the patent office was undemanding, allowing him ample time to focus his mind on his scientific pursuits. Of his time there he said:

> A practical profession is a salvation for a man of my type; an academic career compels a young man to scientific production, and only strong characters can resist the temptation of superficial analysis.

Indeed, the seven years Einstein spent in the patent office were some of his most brilliant, and he would later recall them as the happiest and most fruitful of his life. During this time, he published no less than thirty-two papers, with the pinnacle of his efforts occurring in 1905.

In this year, a twenty-six-year-old Albert Einstein, in his (and perhaps any scientist's) most miraculous scientific year ever, wrote four papers, each having a dramatic impact on physics and also received his PhD in theoretical physics from the University of Zürich.

One of those papers, *On the Movement of Small Particles Suspended in a Stationary Liquid Demanded by the Molecular-Kinetic Theory of Heat*, was a nontrivial extension of his doctoral dissertation work (*On a New*

Determination of Molecular Dimensions). In this work, Einstein considers Brownian motion, which is the motion a "large" particle[158] (Brownian particle) undergoes when suspended[159] in a perfectly still liquid.

The botanist Robert Brown (1773–1858) had observed this type of motion in 1827 while looking through his microscope. Brown noticed that when the very fine particles contained in pollen grains are dispersed in water, they indefinitely undergo an erratic type of motion, as if something is continuously "bumping" into them. Initially, he thought the particles were "alive," consisting of what he called the "primitive molecule" of living matter. However, later he found evidence to the contrary and was never able to explain this motion further.

In his autobiographical notes, Einstein describes his goal to prove the existence of atoms, which is also clear from the opening remarks of his 1905 paper:

> It will be shown in this paper that, according to the molecular-kinetic theory of heat, bodies of microscopically visible size suspended in liquids must, as a result of thermal molecular motions, perform motions of such magnitude that these motions can easily be detected by a microscope.

Einstein believed that the atoms (or molecules) consisting a liquid were always in motion, as described by kinetic theory. Therefore, he felt the motion of one of "Brown's particles" was from the surrounding water molecules constantly colliding with it, which caused the Brownian particle to "bounce" around. But how could this be possible? Even with the poor magnification of Brown's crude microscope, these particles were still large

---

[158] By large I mean something much larger than a single atom or molecule, something the size of a dust particle.

[159] By "suspended in a liquid" we simply mean as opposed to being dissolved in a liquid. For example, if you add sugar or salt to water, they dissolve in the water. Something added to the water that didn't actually dissolve in it but nonetheless sank into the water, where it remained without sinking to the bottom, would be suspended in the liquid.

enough to be seen. Whereas a single water molecule is much too small to be viewed this way,[160] a Brownian particle is much bigger than a water molecule. In fact, if we were to estimate the size difference, taking only the longest length of each for our comparison, we would find that a water molecule is about 6700 times smaller[161] than one of the particles observed by Brown. It's hard to imagine such a tiny water molecule could have any real effect on the much larger Brownian particle – isn't it?

Of course, it isn't only a single molecule that is colliding with the Brownian particle; the Brownian particle sustains collisions in all directions from all the water molecules that surround it. Specifically, groups of water molecules impact the Brownian particle in a collective effort, causing it to finally budge. Einstein imagined the overall motion of the Brownian particle coming from two separate effects: the bouncing motion of the Brownian particle as a result of the water molecule collisions, and the displacement of the particle, whereby it moves from its original position over time.

This *diffusion* of the Brownian particle results from the fact that the total force arising from the colliding water molecules varies in time. This variation means that at one moment in time the Brownian particle is moved a certain amount in one direction, while at another moment in time it's moved by a different amount in another direction. Thus, a "push" here and another "push" thereafter eventually begin to result in not simply a

---

[160] In 1950s, the field ion microscope provided the first technique by which individual atoms could be imaged.

[161] This is a rough estimate, so let's take a moment to understand. A water molecule, as you know, is H_2O. It's a flat molecule where the atoms are connected as H–O–H and the resulting angle is 104.5°. The H–O bond lengths are about 0.1 nanometers (nm), which is 10^{-10} meters (m), or 0.0000000001 m. Also, the distance between the two hydrogens is about 0.15 nm. On the other hand, the size of a Brownian particle is on the order of a few micrometers (μm); one micrometer is 10^{-6} or 0.000001 m. In a letter to his friend, Einstein said that according to his new theory, particles on the order of this size should perform observable movements (via a microscope) due to the thermal motions of the liquid atoms. Therefore, for simplicity I have taken the size of a water molecule as the longest length, being 0.15 nm (the distance between the hydrogen atoms), and then 1 μm for the size of the Brownian particle. This ratio then gives 6666.7 or roughly 6700.

bouncing back-and-forth motion but also in outright movement of the particle from its original position.

Consider a rubber ball that has been dropped to the ground. The ball will (most likely) show two types of motions, where it not only bounces up and down, but also moves away (diffuse) from its initial landing place. This overall motion is similar to that of a Brownian particle. Einstein knew that if one could correctly calculate the amount of movement and it could be verified experimentally, this would prove the existence of atoms. That is to say, the presence of the water molecules would be confirmed as a result of their ability to move the Brownian particle from its starting point.

Einstein derived an expression for the *mean squared displacement* of the Brownian particle.[162] Let's take a moment to understand this quantity. Imagine the Brownian particle as it's being bombarded by the liquid molecules (Figure 12.3). If we watch (using a microscope) the Brownian particle for a certain amount of time (stopwatch in hand), we can determine the total amount it moves from its original position for this specific time interval. What we'll find is that the Brownian particle will travel different distances for different time intervals. Specifically, we'll find, not too surprisingly, that the Brownian particle moves a greater distance given longer time intervals.

Think of the ball analogy again: the ball moves (most likely) farther and farther away from the original position it landed on the floor as the elapsed time becomes longer and longer. So, we watch our Brownian particle for a given time interval and calculate the distance it travels, and then we do it again, and again, and again, until we can calculate a reliable average, or *mean displacement*, for this time interval.

---

[162] These days we determine this using an approach that is somewhat more elegant than Einstein's derivation. One can write down Newton's equation of motion for a Brownian particle that includes a friction term (that is, proportional to the velocity) and a random time-varying force term representing the force on the particle from the liquid (e.g., water molecules). This equation is known in general as a stochastic differential equation because of the latter term and is specifically known as the Langevin equation.

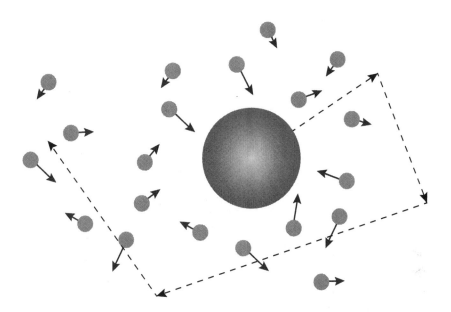

Figure 12.3. A Brownian particle (the larger particle) undergoing collisions with smaller liquid particles. The constant collisions with the smaller liquid particles cause the Brownian particle to be moved. Over time these movements will cause the Brownian particle to diffuse through the liquid (as shown by the dashed arrows).

More specifically we want the *mean squared displacement*. In other words, instead of simply calculating the mean displacement, we square each value of the displacement for a given time interval and calculate the *average* of these squared values. In the end, one has a mean squared displacement for a range of time intervals. Of course, we made it sound pretty easy. In fact, these days you can easily obtain Einstein's predicted value for the mean squared displacement of a Brownian particle using a computer simulation.

Can you imagine, however, having to do the experiment as we more or less just described? That's what Jean Baptiste Perrin did in 1909. Finally, after several failed attempts by other experimentalists who measured values for the mean squared displacement greater than that predicted by Einstein,[163]

---

[163] Einstein's derivation for the mean squared displacement in one dimension was $<x^2> = 2Dt$, where D is the diffusion coefficient and t is the time interval. However, this expression is only valid after a certain amount of time has elapsed. If the time

Perrin confirmed Einstein's result. This finally put an end to the question of the existence of atoms.

Friedrich Wilhelm Ostwald, a strong opponent to the existence of atoms, finally relented in 1909: "I am now convinced that we have recently become possessed of experimental evidence of the discrete or grained nature of matter, which the atomic hypothesis sought in vain for hundreds and thousands of years." However, Ernst Mach, second to none in his opposition of atoms, remained so his whole life.

As an extra bonus, Einstein's equation also provided a new means to determine Avogadro's number.[164] Confirmation of Einstein's prediction gave credibility to kinetic theory, which describes (among other things) atoms and molecules as constantly in motion. Further, Einstein's calculations involved the use of statistical methods such as those introduced by Maxwell and Boltzmann. In fact, he used a more general expression of Boltzmann's equation for entropy in his derivation.

In 1906, Boltzmann returned home to Vienna from his trip to California, totally unaware of Einstein's work; it was later that year that Boltzmann took his own life after years of struggling with bouts of depression. Indeed, Einstein's work would have pleased Boltzmann (as well as Maxwell and Clausius) with its means of confirming the presence of atoms, its validation of kinetic theory, and its statistical approach to the behavior of matter.

Atomic Structure: The Pieces of the Pieces

The atom had always been synonymous with *unchangeable, indestructible, incorruptible,* and *indivisible.* It was envisioned as the smallest portion of

---

interval is too short, the mean squared displacement will follow directly from Newton's equation of motion and go as $<x^2> \sim <v_0^2>t^2$, where v_0 is the average velocity of the system at the given temperature. This, in part, could have led to the error in the initial measurements. Thus, at short times the $<x^2>$ follows from the classical equations of motion, while at longer times, the statistical nature of the system takes over and Einstein's derivation is valid.

[164] This is how Perrin was able to determine the first accurate value as mentioned earlier.

matter from which everything else was made. However, today we know atoms to be made of negatively charged electrons, positively charged protons, and chargeless neutrons. The majority of the mass of an atom is contained in the nucleus, which consists of the neutrons and the protons (with the exception of the hydrogen atom, which contains a single proton as its nucleus). The rest of the atom is made up of electrons, which are tiny in comparison – a single electron is 1836 times lighter than a proton (a proton and neutron are about the same mass).

One of the earlier models of atomic structure envisioned the positively charged nucleus at the center, while the electrons move in orbits around it. There's a lot of empty space in between the electron orbits and the nucleus (it's true that matter consists of a lot of empty space). We'll see in Part IV that this model of the atom provided great insight into its workings, although being fundamentally flawed. Thus, the atom turned out to be even more complicated than originally conceived. Indeed, it's the fundamental particle of matter, but it has an internal structure of its own.

Arriving at a correct description of this structure provided an over-whelming burden to the standard techniques of classical physics, which revealed glaring shortcomings when dealing with particles that are "very small." A correct description of this realm would require a whole new way of conceiving of not only matter, but light as well. Understanding the "unchangeable" atom pretty much changed everything, and it would help usher in a new era of physics and the birth of *quantum mechanics*.

PART IV

Uncertainty: Quantum Mechanics

The more success the quantum theory has, the sillier it looks. How nonphysicists would scoff if they were able to follow the odd course of developments!

– ALBERT EINSTEIN, GERMAN PHYSICIST (1879–1955)

CHAPTER 13

Discrete

Energy's Devious Secret

By 1900, our understanding of energy had come a long way. The first law, established around 1850, ensured that all energy is conserved: never created or destroyed, simply transformed from one form into another.

We now understood that heat was a form of energy, rather than an *imponderable fluid* once known as *caloric*. The existence of atoms as the fundamental constituents of matter was gaining widespread acceptance, helping to reaffirm much-earlier suspicions that heat results from their motion. Indeed, kinetic theory and statistical mechanics obtained results in agreement with experiment by presuming their very existence. Through the work of Maxwell and more so of Boltzmann, the connection between the *macroscopic* world we see around us and the *microscopic* world of atoms not directly observable to us was becoming clearer. Energy's "partner in crime," entropy, was firmly established by the second law. And Boltzmann had shown that entropy's connection to atoms resides in the collection of arrangements, or *microstates*, they explore as they move over time; the more arrangements, or *microstates*, a system of atoms has available to it, the larger its entropy will be. Ah yes, our understanding of energy and its cohorts seemed very complete. Unfortunately, this couldn't have been further from the truth.

Thermal Radiation

When an object is heated, something rather interesting happens. Aside from getting hot, it also emits light, or *thermal radiation*. Most of us are probably familiar with the way a burner of an electric stove changes color as it becomes hotter, starting off as a faint red then becoming redder as the temperature increases. In fact, if we were able to raise the temperature of the burner even higher, we would find the glow of the burner go from red towards more of a blue.

In terms of the *frequency* of the thermal radiation, it shifts from low to high, just like temperature. Although we observe a particular color, a hot object (e.g., stove burner) usually emits thermal radiation over a continuous range of colors, or frequencies, called the object's *frequency spectrum* for the given temperature. In general, any object at a temperature above absolute zero, 0 K (units in Kelvin), will emit radiation.[165] This seemingly innocent nature of materials turned out to be simply impossible to explain according to twentieth-century physics, pointing to the need for a whole new approach.

Our story of thermal radiation and the study of frequency spectra, known as spectroscopy, begins with Gustav Kirchhoff (1824–1887). Kirchhoff had been studying, among other things, the frequency spectrum of the Sun, a very hot object. The spectrum of the Sun had previously been studied by Joseph von Fraunhofer (1787–1826) around 1814. Fraunhofer had noticed that when he passed the light given off by the Sun through a

---

[165] For example, at room temperature, objects emit radiation mostly in the infrared region of the electromagnetic spectrum, although our eyes don't let us directly observe it. However, this is the basis for *thermal imaging* used in some night goggles that allow one to more clearly see objects at night. Another familiar example is an incandescent bulb that is heated (to about 3000 K) to produce visible light. However, in addition to the visible it also emits in infrared and red regions of the electromagnetic spectrum, which constitutes the majority of radiation it gives off.

diffraction grating,[166] it gave a spectrum of colors (like a rainbow)[167] ranging from red to violet, which appeared "mostly continuous" except for the presence of some "black lines," which left the impression that something was missing from the spectrum. Fraunhofer mapped over 570 of these black lines in the Sun's spectrum, denoting the prominent ones as A–K and the weaker ones with other letters. Nonetheless, he was never able to explain their origin.

In a similar experiment, Kirchhoff also noted the presence of these mysterious black lines. However, he noticed something else very interesting. Kirchhoff was collaborating with Robert Wilhelm Eberhard Bunsen (1811–1899) on the spectrum produced by a substance heated by a flame. Bunsen specifically designed a special burner for this purpose, which produced a near colorless flame. By placing a substance into the flame of the *Bunsen burner* and passing a beam of the emitted light through a diffraction grating, a spectrum unique to that particular substance – no spectrum from the flame itself thanks to Bunsen's design – was obtained.

One substance they studied was a sodium salt much like table salt (sodium chloride). When placing the sodium salt in the flame they found that, in addition to the flame turning yellow in color, the emitted light produces a spectrum consisting of two yellow-colored lines upon passing through the diffraction grating, which corresponded precisely with two of those black lines in the Sun's spectrum known as the D lines (according to Fraunhofer's convention).

So, imagine the spectrum of the Sun produced by passing a narrow beam of sunlight through a diffraction grating. After noting this spectrum,

---

[166] As a light ray passes through a diffraction grating, the different colors in it (that is, the light waves of different *frequency* or *wavelength*) are separated out. The resulting collection of colors form a *spectrum*. In this way, a light ray emitted by an atom results in a spectrum that serves as a fingerprint for that given atom type, or element.

[167] A rainbow forms when the Sun comes out right after it rains, or when it's raining while the Sun is out. The water droplets in the air separate the sunlight into the various colors that constitute it: the colors seen in the rainbow. This phenomenon is different from diffraction and involves both refraction and internal reflection of the light with the water droplet.

imagine passing a narrow beam of light emitted from the sodium flame – the beam of sunlight has been blocked now – so that it traces the same path previously followed by the beam of sunlight on its way through the diffraction grating. In this way,[168] Kirchhoff noted that the two yellow lines in the sodium spectrum and the two black D lines of the Sun's spectrum were superimposable. This is clearly intriguing, but the relationship – if one even exists – is not quite clear. But there's more. Kirchhoff returned to the narrow beam of sunlight as before, this time allowing it to pass through the sodium flame prior to the diffraction grating. The Sun's spectrum was as before except that this time the two D lines were darker.

So let's put it all together: the Sun's spectrum alone shows two dark D lines, which are further darkened when the sunlight is first passed through the sodium flame, and the sodium flame's spectrum alone shows two yellow lines that overlap with the two dark D lines present in Sun's spectrum. It's almost as if the sodium flame plays two roles simultaneously: filtering or *absorbing* from the sunlight to make the "presence" – or more correctly, the absence – of the two D lines darker; and producing or *emitting* two D lines of its own that are yellow rather than "black."

In 1859, this series of experiments forced Kirchhoff to a major conclusion: any substance able to emit at a given frequency must also absorb at exactly the same frequency. Moreover, at a given frequency, the emitting power of a substance exactly equals its absorbing power. This is because – and we'll soon understand this more clearly – emission and absorption in atoms occur through the same mechanism, rather than through two separate processes.

An interesting aside is that apparently the atmosphere above the Sun's surface contains, in part, hot sodium atoms that absorb some of the thermal radiation emanating from the Sun's core, resulting in the two dark D lines, which become even darker after passing through a sodium flame. Thus, these experiments also reveal something about an atmosphere some ninety-three million miles away from Earth.

---

[168] To be sure, I am only offering a very plausible layout of how Kirchhoff may have performed his experiment.

Kirchhoff went on to formulate a relationship between the rate of energy emitted and the amount absorbed by an object at a particular frequency when the object is in *thermal equilibrium*. *Kirchhoff's Law* and the challenge he posed as a consequence of it drove theorists and experimentalists alike in a revolution of physics that would eventually culminate in *quantum mechanics*. Perhaps we should take a closer look at this instigative theory.

To a reasonable approximation, when you place food into an oven and heat it to a specific temperature, once that temperature is reached and thereafter maintained, the system (oven and food) will eventually achieve thermal equilibrium. So let's say we place a turkey in the oven and set the oven to 400°F. Once it has reached 400°F and the temperature is maintained awhile, we can say (for our purposes) that the oven and the turkey are at thermal equilibrium.

Further, from our previous discussion, we expect the turkey and the inside walls of the oven to be emitting and absorbing thermal radiation. Now that's not to say that the turkey will take on a red glow like an electric stove burner; they are different materials after all. Nonetheless, it's emitting and absorbing thermal radiation – mostly in the infrared region, which we can't see with our eyes alone. Since thermal equilibrium has been established, the turkey is emitting and absorbing this thermal radiation at the same rate. This also goes for the inside walls of the oven and anything else that we add and allow to attain thermal equilibrium. So let's add a baked potato and bring it to thermal equilibrium. Let's focus on just the turkey and the potato.

To simplify things a bit, and to focus our attention on the turkey and the potato – the things *inside* the oven – let's replace the oven with a "box" that is able to maintain everything inside at our desired temperature. Moreover, let's design the walls such that they neither absorb nor emit thermal radiation of their own; they perfectly reflect the radiation of the turkey and potato. Therefore, the thermal radiation being emitted and absorbed only involves the turkey and the potato. Understanding Kirchhoff's Law requires us to understand two simple things:

- A *portion of* the thermal radiation emitted from the turkey accounts for the *total amount* of thermal radiation being absorbed by the potato, and vice versa.[169]
- As a consequence of thermal equilibrium, both the turkey and the potato will emit thermal radiation at the same rate they absorb it.[170]

Written mathematically, these concepts mean:[171]

$$\frac{E_1}{\alpha_1} = \frac{E_2}{\alpha_2}$$

where E is amount of energy per time emitted (it's a rate or the *emissive power*) by an object, α is the fraction of energy absorbed by an object, and the subscripts label the object (e.g., turkey = 1, and potato = 2).

Let's imagine an ideal object that absorbs all the thermal radiation that hits it regardless of the frequency. This means $\alpha = 1$, as in 100% of the radiation that hits the ideal object is absorbed. Such a body would be black in color, a *blackbody*. As we already know, an object that absorbs at a given frequency must also emit at this frequency (under the appropriate conditions, such as being heated). In the case of our blackbody, we insist

---

[169] You could argue that the potato reabsorbs some of the thermal radiation it emits, therefore the total amount of thermal radiation absorbed is a fraction of the amount emitted from the turkey plus the amount the potato reabsorbs of its own. I've chosen to ignore the possibility of reabsorption here since the final mathematical relations I want to discuss won't change.

[170] The second law requires that each object in the system (e.g., turkey and potato) individually emits and absorbs thermal radiation at the same rate. The first law merely requires that the *total* rate emitted and *total* rate absorbed equal each other. However, only adhering to the first law would allow for the scenario that (in a system of two objects) one object only emits while the other only absorbs. Thus, the emitting object continues to cool while the absorbing object continues to warm. In other words, we have a colder object transferring heat to a warmer object, which is a strict violation of the second law. This disaster is avoided by concluding that each object must emit and absorb at the same rate when the system is at equilibrium.

[171] Amazingly, this equation doesn't depend on the shape or composition of the objects or the box.

that when heated (just like when the sodium was placed in the flame) it will emit its spectrum – at all frequencies.

Therefore, if one of the objects in our system is a blackbody – say we swap out the potato – then our equation reduces to:

$$\frac{E_1}{\alpha_1} = E_{\text{blackbody}}$$

and since the exact identity of the other object doesn't matter (turkey, potato, etc.), we can drop the subscripts on the left-hand side and simply write:

$$\frac{E}{\alpha} = E_{\text{blackbody}}$$

which in plain words means: the power of the thermal radiation an object emits (emissive power) over the fraction of thermal radiation it absorbs (this ratio) is equal to the emissive power of an ideal object, called a blackbody, that emits and absorbs at all frequencies. Usually this equation is written in terms of a given frequency v, and from our discussion it's clear we mean at a specific temperature T:

$$\frac{E}{\alpha} = E_{\text{blackbody}}(v, T)$$

The significance of this equation is that knowing the spectrum of the blackbody (the right-hand side of this equation) gives us the ratio (left-hand side) for any object. In other words, for any object in thermal equilibrium at temperature T, the ratio of the emissive power over the fraction of thermal radiation absorbed at a certain frequency v is a *universal function* equal to the corresponding emissive power of the blackbody at that frequency when in thermal equilibrium at that temperature. Although a blackbody is an idealized object, it's essentially realized by certain materials in nature, such as soot or graphite.

Kirchhoff challenged theorists and experimentalists to find the form of the blackbody emission spectrum, as he considered it of fundamental importance:

It is a highly important task to find [the blackbody emission spectrum]. Great difficulties stand in the way of its experimental determination; nevertheless, there appear grounds for the hope that it can be found by experiments because there is no doubt that it has a simple form, as do all functions which do not depend on the properties of individual bodies and which one has become acquainted with before now.

Kirchhoff's anticipation of the experimental challenges turned out to be right. It would be almost forty years before enough experimental evidence was collected to allow an answer to Kirchhoff's challenge. In 1900, a forty-two-year-old Max Planck finally put Kirchhoff's challenge to rest.

The Blackbody Radiation Challenge

Max Planck (1858–1947) was born in Kiel (in modern-day Germany), the sixth child[172] to the distinguished jurist and professor of law at the University of Kiel, Johann Julius Wilhelm Planck, and his second wife, Emma Patzig. His family culture would bestow in Planck's life and work a sense of excellence in scholarship, incorruptibility, idealism, reliability, and generosity.

In 1867, when Planck was nine, his father received an appointment at the University of Munich. The family moved, and Planck enrolled in the city's Maximilian Gymnasium, where his interest in physics and mathematics was piqued. However, Planck excelled in his other studies as well, in particular music. Thus, at the time of graduation, now sixteen, Planck had the difficult decision of choosing a future in either music or physics. He chose physics, but music would remain an important part of his life. Indeed, Planck was an excellent pianist (with the gift of absolute pitch) who found daily pleasure in playing the works of Schubert and Brahms.

In 1874, Planck entered the University of Munich intending to study physics. Professor of physics, Philipp von Jolly (1809–1884), warned Planck

---

[172] Two of his siblings were from his father's first marriage.

against such a pursuit: "In this field, almost everything is already discovered, and all that remains is to fill a few holes."

Wisely, Planck ignored this advice. Under Jolly's supervision, Planck performed the only experiments of his career, eventually transitioning to theoretical physics. In 1877, Planck studied for a year at the University of Berlin under two of Germany's most prominent physicists, Hermann Helmholtz (recall from Part I) and Gustav Kirchhoff. Unfortunately, neither offered much inspiration to Planck in the lecture hall. Planck recalled Helmholtz as poorly prepared and Kirchhoff as "dry and monotonous." However, through independent study of their and Clausius' works on thermodynamics (see Part II), Planck found what he had been searching for at last.

Planck approached physics with a fervor – or a "hunger of the soul" as Einstein described it – in a search for things absolute and fundamental. He said:

> it is of paramount importance that the outside world is something independent from man, something absolute, and the quest for the laws which apply to this absolute appeared to me as the most sublime scientific pursuit in life.

Indeed, Planck saw the search for the absolute "as the noblest and most worthwhile task of science."

In 1879, Planck, now twenty-one, received his PhD for his thesis on the second law and the entropy concept. Thermodynamics, especially the entropy concept, would remain a central theme throughout most of Planck's work. Unfortunately, Planck's career was off to a slow start, as he was unable to make much of an impression with his work on the entropy concept.

One reason for this might simply have been that this area of study was still relatively new. More disappointment followed when Planck discovered that much of his work on entropy theory had already been done by Josiah Willard Gibbs (1839–1903). Moreover, his work on the thermodynamics of dilute solutions lacked the chemical insight so successfully applied by others such as Jacobus Henricus van 't Hoff (1852–1911). However, his efforts weren't in vain and the skills that he'd been refining were finally going to serve him well.

Although proposals for the form of the blackbody spectrum started showing up in the 1860s, real progress followed the experimental successes in the 1890s.[173] In 1893, Wilhelm Wien (1864–1928) provided a generalized mathematical form for the blackbody spectrum, known as *Wien's Displacement Law*. In 1896, he went further by giving a more specific mathematical form to his previous one. This version, *Wien's Radiation Law*, showed excellent agreement with the available experimental data, thus appearing to be the solution long sought after to Kirchhoff's challenge – or so it seemed.

Between 1887 and 1894, Planck had been spending much of his time on the exciting new field of physical chemistry pioneered by Svante August Arrhenius (1859–1927) and van 't Hoff. As already mentioned, this work went mostly unnoticed, a fact Planck would later recall. Nonetheless, it undoubtedly helped to secure his position as special chair in mathematical physics at the University of Kiel in 1885,[174] and subsequently in 1889 as Kirchhoff's successor at the University of Berlin,[175] becoming a full professor in 1892.

Planck's move to Berlin brought him to what would soon become the epicenter for theoretical and experimental research on blackbody radiation. By the 1890s, hardly a physicist in Berlin was unaware that Kirchhoff, Boltzmann, and Wien had secured a role for thermodynamics in the solution of the blackbody problem. Thus, in 1894 when Planck

---

[173] It's worth mentioning that in 1879, Josef Stefan (1835–1893) proposed, based on experimental data, that the total energy radiated by a hot object goes as T^4. In general, this statement is not true, and in 1884 Boltzmann derived the precise result showing that the T^4 variation speculated by Stefan only applies to a blackbody, not just any hot object.

[174] In his 1848 *Scientific Autobiography*, Planck notes that the close friendship his father had with the professor of physics at the University of Kiel played a role in his appointment.

[175] Helmholtz might have been helpful in Planck obtaining this position.

embarked on the blackbody problem, he expected his beloved tools of thermodynamics – in particular the entropy concept, which he had been polishing for some time on physical chemistry problems – to serve him well.

Several things enticed Planck about the blackbody problem. First, Planck was interested in securing the role of thermodynamics and the entropy concept in the field of electrodynamics (the study of electricity, magnetism, and light, e.g., thermal radiation). Further, this was a high-profile problem in the physics community, and evidently Planck often discussed the blackbody problem with his Berlin colleagues, in particular Wien and Heinrich Rubens (1865–1922), both of whom were very involved in blackbody research. But most of all, for Planck the blackbody problem provided the opportunity to seek out something absolute:

By the measurements of [Otto] Lummer and [Ernst] Pringsheim at the Physikalisch-Technische Reichsanstalt, performed to investigate the spectrum of heat radiation, my attention was directed to the theorem of Kirchhoff, stating that in an evacuated cavity – bound by totally reflecting walls and containing emitting and absorbing bodies that are completely arbitrary – a state will be established in course of time, in which all bodies assume the same temperature and in which all properties of the radiation – even its [thermal radiation spectrum] – do not depend on the structure and composition of the bodies, but solely on the temperature. This [thermal radiation spectrum], therefore, represents an absolute quantity; and, since the search for the absolute always appeared to me to be the most beautiful ("schönste") task of research, I eagerly started to deal with it.

Planck had one last motive for working on the blackbody radiation problem. He had been commissioned by several electricity companies to develop light bulbs that would provide the most light output for the least amount of energy used.

Planck imagined a box with perfectly reflecting walls containing objects that he called "resonators."[176] There are many of these resonators, with each one emitting and absorbing thermal radiation at various frequencies. Today, we would liken Planck's resonator to an atom or molecule. However, there was no need for Planck to provide any information on the composition of the resonator since – as we know – Kirchhoff's Law holds regardless of the specific details of the objects in the system.

Indeed, Planck maintained this generality throughout his approach and leveraged it to his advantage every step of the way. In this way, Planck's style has much in common with Fourier's in the development of his law (see Part II): they both focused on the general details of the system at hand rather than getting bogged down by the microscopic details.

At that time, Planck wasn't a believer in the existence of atoms and molecules. Indeed, on this point, he was one of Boltzmann's biggest opponents, explicitly rejecting atoms in 1881:

> When correctly used, the [second law of thermodynamics] is incompatible with the hypothesis of finite atoms. One should therefore expect that in the course of the further development of the theory, there will be a fight between these two hypotheses that will cost the life of one of them. It would be premature to predict the outcome of this fight now; but for the moment it seems to me that, in spite of the great successes of the atomic theory in the past, we will finally have to give it up ....

The main issue for Planck was that the concept of atoms and molecules, and the theories that made use of them, like kinetic theory and statistical mechanics, undermined his interpretation of entropy. Planck's vision of entropy was directly opposed to Boltzmann's microscopic version with its atoms, microstates, and – most of all – its probabilities.

---

[176] Throughout his study, Planck would make use of this resonator concept, sometimes considering a collection of resonators, and sometimes considering a single resonator only.

For Planck, entropy was an "absolute" law, not one involving probabilistic or statistical behavior as Boltzmann had described it. Specifically, for Planck, a system's approach to equilibrium from a nonequilibrium state (more specifically *macrostate* as discussed in Part II) not only had to result in an increase in entropy – as demanded by the second law – but it had to happen *every step of the way*.

Boltzmann had already learned his lesson in 1876 when a criticism by Loschmidt caused him to reconsider the nature of the second law. By 1877, Boltzmann had concluded that the nature of the second law was inherently probabilistic. As a result, Boltzmann viewed the increase in entropy (demanded by the second law) a system incurs as it approaches equilibrium from a nonequilibrium state as an *overall* increase, not one occurring at every single step along the way.

Thus, a system can "take two steps forward and one step back," so to speak, in the sense that it's possible for a system to move from a state of higher entropy to one of lower entropy and so on, as long as the *overall* entropy increases by the time it reaches equilibrium. Nonetheless, Planck was in good company with his mistake.

In 1872, when he wrote *Further Studies on the Thermal Equilibrium of Gas Molecules*, Boltzmann not only held the same view as Planck, but was convinced he had actually proved it:

> our result is equivalent to a proof that the entropy must always *increase* or remain constant [when at equilibrium], and thus provides a microscopic interpretation of the second law of thermodynamics.

And in 1903, a young (twenty-three-year-old) Einstein also fell victim to this same mistake, writing in his only publication that year, *A Theory of the Foundations of Thermodynamics*:

> We will have to assume that more probable distributions will always follow less probable ones, that is, that W [or entropy] *always* increases until the distribution becomes constant and W [or entropy] has reached a maximum. [my italics]

He makes this mistake once again in 1904, in his publication *On the General Molecular Theory of Heat*. Einstein had been studying Boltzmann's book since 1901, but apparently missed the slightly buried discussion on Loschmidt's objection that prompted Boltzmann's own reflections on the topic. Finally, in a 1910 publication, Einstein correctly states the second law. In a nutshell, the mistake Boltzmann, Einstein, and Planck all made was in thinking that the entropy of a system moving from nonequilibrium to equilibrium *always* increases at every step along the way. It's true that in terms of the probability, the system *most likely* increases its entropy at every step along the way, but *not always*. Indeed, a main motivation for Planck's tackling of the blackbody problem was to prove that the entropy *always increases*, and it greatly influenced the way he chose to solve the problem.

With his resonator model in hand, Planck was ready. We imagine Planck's system in a nonequilibrium state at some initial temperature, with the resonators absorbing and emitting thermal radiation at their various frequencies. As time goes by, the temperature will change, and the entropy will increase irreversibly (see Part II). Eventually, the system of resonators will come to equilibrium: the temperature will be constant and the entropy will be at its maximum value.

Planck wanted to develop a theory describing the state of the system over time as it evolved from a nonequilibrium to equilibrium, and in the process show that the resulting spectrum of the thermal radiation emitted by the resonators could be given by none other than the blackbody spectrum. Moreover, Planck wanted a theory without recourse to probabilities and statistics, as it would tarnish the absolute nature of his vision of entropy.

Planck started his search in 1894, making substantial progress that convinced him he could be successful with an approach devoid of a probabilistic or a statistical interpretation of entropy. However, by 1898, after making several concessions to Boltzmann's criticisms, it became clear to him that he needed a different approach.

Planck's system of resonators interacting with thermal radiation in the confines of a box with perfectly reflecting walls isn't the only system to evolve irreversibly to equilibrium from an initial state of nonequilibrium.[177] Consider a system of gas atoms at nonequilibrium. Such a system will be driven to equilibrium through their collisions with each other. The evolution will be irreversible, with the entropy increasing along the way. Once at equilibrium, the system will be at maximum entropy, having achieved (in the case of an ideal gas) a Maxwell distribution of velocities, which will thereafter be maintained. Indeed, in 1872 Boltzmann proved all this in *Further Studies on the Thermal Equilibrium of Gas Molecules*. He had advised that Planck use a similar approach for his problem. After much resistance to Boltzmann's advice, by the spring of 1898 Planck was ready to concede.

It wasn't that Planck hadn't made progress. On the contrary: by this time, his almost four years of work produced an equation describing the dynamics of a single oscillator interacting with the thermal radiation. By following a regimen similar to the one Boltzmann did in 1872, Planck was able to extend this equation to a more "fundamental equation," relating the average energy of a resonator to the general form of the blackbody spectrum for the system at equilibrium; he essentially had the equation for matter in equilibrium with radiation.

However, whereas Boltzmann was able to arrive at an equation governing the dynamics of a system of gas atoms evolving from nonequilibrium to equilibrium, Planck, although using a similar approach to Boltzmann's, was unable to produce such a relationship for the evolution of the thermal radiation emitted by the resonators. Thus, Planck had to abandon his hope of proving the absolute nature of entropy's increase, although he maintained his belief in its existence for fifteen more years. However, the greatest disappointment must have been that none of his efforts resulted in an explicit form for the equilibrium blackbody spectrum. At this point, Planck again turned to his old friend, entropy.

Using his fundamental equation and Wien's Displacement Law, Planck was able to arrive at an equation for the entropy of a resonator. From here, he made easy work of deriving Wien's Radiation Law. Wien had proposed

---

[177] In fact, every nonequilibrium system that is tending towards equilibrium will do so irreversibly.

his radiation law based on some very shaky arguments, never providing a rigorous derivation. Therefore, Planck was the first to provide an actual derivation of it, which became known as the *Wien-Planck Law*.

In early 1899, at the time of Planck's fifth publication, recounting his efforts, the Wien-Planck Law was in excellent agreement with experimental measurements. Although falling short of his overall goal, Planck was convinced he'd successfully derived the universal function for the blackbody spectrum that Kirchhoff had challenged physicists to find some forty years earlier. However, his victory would be short-lived.

Discrepancy and Desperation

By the spring of 1900, improved experimental techniques revealed new results showing discrepancy with the Wien-Planck Law. It was beginning to look as if Planck's almost six years of effort had been in vain. Planck decided the error was in his derivation of the entropy of the resonator. Regrouping, Planck proposed a new form for the entropy of a resonator, and from it he was able to obtain a new expression for the blackbody spectrum. This time he gets it right, and the new result is shown to be in perfect agreement with all the new experimental results.

Planck could've rested on his laurels at this point, probably winning the Nobel Prize for his efforts, either independently or sharing it with Wien. However, Planck needed to understand the "absolute" nature of his new expression. After all, at this point Planck had produced – based on a fair amount of "inspired guessing" – little more than a powerful equation relating the interaction of matter and radiation, without much physical insight into its true nature. The question is this: What is it about the interaction of matter and radiation that leads to Planck's form of the blackbody spectrum?

To answer this, Planck again appeals to Boltzmann, this time his work of 1877. Previously a resilient opponent of Boltzmann's statistical interpretation of entropy, Planck seemed to be coming around a bit:

> I conjectured that it would have to be possible to calculate this quantity [the entropy of a resonator] by introducing probabilistic

considerations, whose importance for the second law of thermodynamics was first revealed by L. Boltzmann.

Although making concessions to Boltzmann, he's not a total convert to Boltzmann's statistical approach; Planck has his own interpretation as to the nature of statistics in physical phenomena. He also doesn't follow Boltzmann's original approach all that closely, taking a few liberties as needed. Planck would later (in 1931) refer to this part of his program as "an act of desperation," explaining that he "had to obtain a positive result, under any circumstances and at whatever cost."

The starting point for Planck's "act of desperation" was Boltzmann's equation, introduced in Part II:

$$S = k \ln W$$

Recall that this expression describes the total entropy of the system in a given macrostate, or physical state, for the corresponding total number of microstates, W. Actually, Boltzmann originally wrote this equation in the form of a proportionality:

$$S \propto \ln W$$

rather than the equality we know it as today. It was Planck who later included and emphasized the importance of the constant, k, which we now call Boltzmann's constant. Further, Boltzmann proposed a method to calculate the total number of microstates, W, or *complexions* as Boltzmann called them.

It was an improved version of his 1868 method and was intended to be a complementary approach to his work of 1872. Boltzmann considered a system of N gas atoms with a constant total energy, E. A microstate of such a system would describe each of the N gas atoms as having a certain portion of the total energy. To enable such a description, Boltzmann imagined the total energy as being divided up as

$$E = P\varepsilon$$

In this way, the total energy of the system would come in "chunks" of ε, and a single gas atom could have an energy of 0, ε, 2ε, 3ε, ... $P\varepsilon$.

Imagine a system comprising two atoms ($N = 2$), where the energy is divided into three chunks ($P = 3$), so that the possible energies for a given atom are then 0, ε, 2ε, 3ε. The only possible microstates – given the requirement that the energy over each atom must total that of the system ($E = 3\varepsilon$) – would be (0, 3ε), (3ε, 0), (2ε, ε), (ε, 2ε), where the first energy in the parenthesis refers to "atom 1," and the second refers to "atom 2." Each microstate labels the atom (either 1 or 2) along with its corresponding value of the energy. In other words, we have made both the energy and the atoms *distinguishable* (we'll have more to say about this later). For an arbitrary number of gas atoms and chunks of energy (any value of N and P), one can write down a general expression for W,[178] and therefore the entropy.

For Boltzmann this approach was simply a means to an end. In a final step, he allowed the system size to approach an extremely large number of atoms. Further, he made those chunks of energy extremely small. We've already seen that real systems, like a balloon filled with gas atoms, contain an extremely large number of atoms, which is why Boltzmann considered a system of such size; he wanted to make a connection to real physical systems. As for allowing the chunks of energy to become extremely small, well, that too was motivated by real physical considerations.

Boltzmann considered the energy of a system as being distributed in chunks a useful bit of fiction, a mathematical trick that made it possible to write an expression for entropy. He explains:

> This fiction [the chunks of energy] does not, to be sure, correspond to any realizable [physical] problem, but it is indeed a problem

---

[178] The expression for W is in the form of a *combinatorial* given by:
$$W = \frac{(N-1+P)!}{P!(N-1)!}$$
In Boltzmann's hands this was the total number of different ways P "chunks" of *distinguishable* energy could be spread over a system of N *distinguishable* gas atoms, which was the total number of microstates for the system.

which is much easier to handle mathematically and which goes over directly into the [physical] problem to be solved ....

However, in Boltzmann's mind, and in pretty much every other scientist's mind at the time, energy didn't behave this way. Rather, energy was considered to be *continuous*, not *discontinuous* or composed of chunks. However, Planck had his own take on *Boltzmann's complexion method*.

By November 1900, realizing that his inspired guess needed a stronger foundation, Planck once again turned to Boltzmann's statistical interpretation of entropy and the method he used to calculate it. To make a connection to Boltzmann's system of N gas atoms with a constant total energy E Planck considers a collection of N resonators – rather than a single resonator – that are all at the same frequency v with their own constant total energy.

The expression Planck derived for entropy was for that of a single resonator at a given frequency v, not a collection of N. Thus, Planck proposes to simply divide Boltzmann's equation ($S = k \ln W$) by N to convert it to the entropy for a single resonator.[179] Now Planck has two expressions for the entropy of a resonator: the one he derived himself and the new expression obtained by the simple division of Boltzmann's equation. Further, like the entropy of the system, Planck also considered the total energy of the system to be the sum of the energies of all the resonators.

Planck's final steps are straightforward: use Boltzmann's method to obtain W, which will complete the modified expression of Boltzmann's equation; equate this version of the entropy to the one Planck obtained himself; and then hope for some physical insight. Planck proceeded to implement Boltzmann's "mathematical trick" of dividing the total energy of the system into P "chunks" of ε to be spread over the N resonators. Of this he says:

---

[179] This amounts to saying the system consists of N *independent* resonators. In this case, the total entropy of such a system would be the sum of the entropies of all the resonators, or $N \times$ (entropy of a single resonator). Moreover, this analogy is perfectly accurate since Planck considered his system at equilibrium. As such, a system of N independent resonators at equilibrium should be comparable to a system of just one of those resonators at equilibrium.

> If E [the total energy of the system of N resonators] is regarded as infinitely divisible, an infinite number of different distributions is possible. We, however, consider – and this is the essential point – E to be composed of a determinate number of equal finite parts ….

Now, with the energy in chunks as required, Planck calculates the total number of microstates (W) for his system of N resonators and the corresponding entropy, and he finishes by setting his expression of entropy (obtained with Boltzmann's method) equal to his own. He finds something very interesting. In short, in order for the two expressions of entropy to be equal, he simply requires that those little chunks of energy equal:

$$\varepsilon = h\nu$$

where he introduces another (in addition to k) constant, h, which we now call Planck's constant.

Thus, unlike Boltzmann, who considered the chunks of energy a convenient tool and eliminated them in the end by making them extremely small, Planck's theory demands their existence. And what was the physical insight Planck so desperately sought? Evidently, energy *really* does come in discontinuous chunks! In other words, in terms of one of Planck's resonators (which today we would liken to an atom or molecule), the allowed energy values are *discrete* (0, ε, 2ε, 3ε, etc.) rather than a continuous distribution. Moreover, if a resonator's energy goes up or down during its interaction with light, it must do so in increments of ε, no more, no less.

In general, then, the energy of a resonator has the form:

$$E_{\text{resonator}} = mh\nu$$

where $m = 0, 1, 2, 3, \ldots$.

Planck presented his derivation on December 14, 1900. He had been victorious in his search for the blackbody spectrum. Further, he had gained his desired physical insight into the interaction between matter and radiation. Nonetheless, the price he paid was quite high.

He had to appeal to Boltzmann's method of obtaining the total number of microstates in order to obtain an expression for entropy. Boltzmann's method wasn't mainstream at all and was considered by most to be questionable at best. Worse, unlike Boltzmann – who was able to neatly dispose of those annoying chunks of energy in the end – Planck was stuck with them for good since their elimination resulted in complete and utter failure of his theory.

A full acceptance of this rather devious character of energy meant that all of physics, as Planck had known it, would drastically change forever. Understandably, Planck was reluctant in his new role as a revolutionary, and he did very little to promote the discontinuous, or discrete, nature of energy, which today we call energy *quanta* (or *quantum* for a single "chunk").

He held to the idea that quanta of energy were a mathematical artifact and hoped that further refinements to his theory would lead back to the "old familiar physics" (classical physics) with less drastic results. He – and pretty much everyone else – chose to focus on the remarkable accuracy of *Planck's Radiation Law*, rather than the annoying energy quanta that it implied. It took almost eight years after he first presented his *quantum theory* of discrete energy before Planck could finally admit it represented the fundamental nature of energy:

> there exists a certain threshold: the resonator does not respond at all to very small excitations; if it responds to larger ones, it does so only in such a way that its energy is an integral multiple of the energy element hv, so that the instantaneous value of the energy is always represented by such an integral multiple.

While Planck and others might have been hesitant in their acceptance of the energy quanta, there was one who embraced it almost immediately.

CHAPTER 14

Light Quanta

Particles and Waves: The Beginning

In 1905, at the age of twenty-six, Einstein published four major papers and finished his PhD thesis. Each of these papers was groundbreaking and would change physics forever. However, it was the first of these, *On a Heuristic Point of View Concerning the Production and Transformation of Light*, which Einstein referred to as "very revolutionary"[180] – the only time he would ever say this about any of his work, in fact – that would, in part, win him the Nobel Prize in 1921.[181]

Indeed, Einstein's view on light was very revolutionary and it would be almost twenty years before it would be neatly worked into physics. Unlike Planck, Einstein was very comfortable with using statistical approaches (such as kinetic theory and statistical mechanics) to solve physical problems. In fact, Einstein spent a considerable amount of time with these methods.

In a series of three publications appearing between 1902 and 1904, Einstein independently reintroduced some of the ideas of statistical

---

[180] During the spring of 1905, Einstein wrote to his old school friend (from the University of Zürich), Conrad Habicht, about this work with quite some enthusiasm.

[181] However, Einstein actually received it in 1922.

mechanics that had already been formulated by Boltzmann and Gibbs. Apparently, Einstein was somewhat aware of Boltzmann's works, but undoubtedly completely unaware of Gibbs'. It seems Einstein's knowledge of Boltzmann's work came from his *Lectures on Gas Theory*, a two-volume work appearing in 1896 and 1898. This is rather unfortunate since this work isn't intended as an overview of Boltzmann's early work. In particular, Boltzmann's complexion method of 1877, which Planck used in his derivation of energy quanta, is mentioned only in passing, with the (most likely confused) reader being directed to the original reference, which it seems Einstein wasn't able to obtain.

Moreover, whereas Boltzmann was an outstanding lecturer, his writing was often unnecessarily long and dense, with the main conclusions often securely hidden among a forest of calculations. Maxwell said of Boltzmann's writings: "By the study of Boltzmann I have been unable to understand him. He could not understand me on account of my shortness, and his length was and is an equal stumbling block to me."

Perhaps this lack of familiarity served Einstein well, as it gave him the opportunity to develop statistical mechanics "from scratch," truly making it his own in those early years of his physics career. As will become clear, Einstein's approach to studying quanta, or quantum theory, was statistical mechanical in nature, and the tools he developed in those formative years would serve him well for more than twenty years in his pursuits, especially in endeavors concerning the nature of light.

Entropy and the Hypothesis

As it did in Planck's work, entropy also played a major role in Einstein's 1905 paper. Einstein was interested in calculating the entropy of a system consisting of light in a box with volume V_0. While it may seem strange to consider a box full of light, the system itself isn't much different from others we've discussed, such as a box (or balloon) of gas atoms.

Starting with an equation (originally derived by Wien) and using Wien's Radiation Law (not Planck's), Einstein obtained an explicit form for the entropy, S_0, of the light in a box. He then went on to consider the entropy,

S, that would result from confining the light to a smaller volume, V (a *subvolume*), within the box. For this difference he found:

$$S - S_0 = k \ln \left(\frac{V}{V_0} \right)^{E/h\nu}$$

where E is the total energy of the light in the box,[182] and $h\nu$ – as in Planck's equation – is the energy quanta.

Now, if we identify W of Boltzmann's equation[183] with:

$$W = \left(\frac{V}{V_0} \right)^{E/h\nu}$$

of Einstein's equation, we then have:

$$S = k \ln W + S_0$$

The extra term, S_0, results from the fact that we are imagining the system as, say, being in another macrostate prior to the present one being considered with entropy, S. This is a more general form of Boltzmann's equation, which originally appeared in a paper by Planck.

Now, Einstein considered the same change in entropy for a system containing N ideal gas atoms.[184] Ideal gas atoms don't interact with, or rather

---

[182] It's interesting that the energy of the light in the box doesn't change with the volume change of the box. In Einstein's equations this comes from the fact that he is considering the frequency of light only within a small range, and this range doesn't change when the volume change occurs. Since the frequency range of the light doesn't change when the volume changes, neither does its total energy.

[183] Boltzmann's equation for entropy assumes that the probability of a given microstate occurring is the same for all microstates. In general, this isn't true, and usually a more complicated expression is needed to calculate the entropy.

[184] Interestingly, even though the expression for the entropy difference of this system was (and still is) well known, Einstein spent about two pages re-deriving it in his own personal style. Just prior to proceeding with this derivation, Einstein took a moment to express what seems to be his displeasure in the way the determination of W had been handled by others. Specifically, he felt the counting method introduced

behave independently of, each other. Moreover, here we are talking about the *classical* ideal gas, whereas later we'll discuss its *quantum* analog. Once again, Einstein considered the system, originally in a box of volume V_0, to suddenly become confined to the subvolume V.

$$S - S_0 = k \ln\left(\frac{V}{V_0}\right)^N$$

where, in this case:

$$W = \left(\frac{V}{V_0}\right)^N$$

Physically, when it comes to a change in entropy (due to a volume change), these equations are telling us that the two systems behave in the same manner. When we compare them (or more simply their W terms), we see the only difference is between the N (of the ideal gas version) and the $E/h\nu$ (of the light version).

In other words, where we have N ideal gas atoms in one, in the other we have the total energy of light, E, divided into "chunks" of $h\nu$. Einstein made a point of considering only radiation at low density (or low intensity) according to Wien's Radiation Law, not making use of Planck's version at all.[185] Moreover, a real gas at low density is well described in terms of an ideal gas.

With these considerations and the equations, Einstein had constructed an ingenious analogy between these two systems. But what are those chunks of $h\nu$ spread over in the box full of light? Einstein says,

---

by Boltzmann and implemented by Planck was rather contrived. In fact, he promised to introduce a new method of his own in a subsequent paper, "and I hope that this will remove a logical difficulty that still hinders the implementation of Boltzmann's principle [equation]." The paper Einstein promised never came. Clearly, part (if not most) of his motivation for spending time on the re-derivation was to motivate this future paper. Perhaps Einstein had envisioned a means of extending the determination of W one obtains for an ideal gas to other systems. Unfortunately, his optimism was misplaced, as the ideal gas presents a special case in which W is easily calculated.

[185] It's often mistakenly said that Einstein was building on Planck's work with this work. This is untrue, as Einstein makes no use of Planck's work at all here.

> Monochromatic radiation of low density (within the range of validity of Wien's radiation formula) behaves thermodynamically as if it consisted of mutually independent energy quanta of magnitude [hv].

In other words, according to Einstein, light of a given frequency, v (at low density), behaves as if it's made up of chunks of hv – light consists of *light quanta*. That is, the energy of a quantum of light is:

$$E_{\text{quantum of light}} = hv$$

Recall that for Planck, these chunks represented the pieces of energy, or *energy quanta*, that were spread among his N resonators. However, Planck's system was light in equilibrium with the N resonators, where the resonators played the role of matter, whereas Einstein's system was light in equilibrium with itself. For Planck, those chunks had nothing to do with light and were exclusive to the resonators. In other words, while Planck required a resonator to exchange energy in chunks, the light it was passing this energy back and forth with didn't have this requirement – a bit puzzling. Then Einstein goes on to say:

> If, with regard to the dependence of its entropy on volume, a monochromatic radiation (of sufficiently low density) behaves like a discontinuous medium consisting of energy quanta of magnitude [hv], then it seems reasonable to investigate whether the laws of generation and conversion of light are also so constituted as if light consisted of such energy quanta.

Einstein is saying that if light behaves as light quanta when light is by itself, then perhaps it also behaves this way when interacting with matter. To this end, he considered known experimental observations and showed how they could be interpreted in terms of his newly formed light quanta hypothesis. The most notable of these phenomena he discussed was the *photoelectric effect*.

The Way It Was: Before Light Quanta

The nature of light had been debated many times before Einstein revisited the topic in 1905. As with matter, the earliest theories of light came from the Greeks. Democritus, the original atomist, believed light consisted of particles, not unlike the atoms he envisioned all matter to be composed of. Aristotle believed light to be a "disturbance" in – what he considered as one of his four elements – air. Descartes, in his 1637 publication *Optics*, described light as a "lively action" that passes through the air and other transparent bodies.[186] For the most part, Aristotle's "disturbance" and Descartes "lively action" were very early descriptions of light as a sort of *wave*, not that unlike the waves of the ocean.

In 1666, Newton bought his first prism[187] with the motivation of disproving Descartes' wave theory of light. In 1672, he gave a brief account of his findings, in the form of a letter, to the Royal Society, and, after a bit of convincing, it was published in the *Philosophical Transactions of the Royal Society*. Newton's work met with much praise, although not unanimous. However, one critic's words would resound with Newton, thus beginning a lifetime feud.

Robert Hooke, who was considered the expert on the subject of light in England, sent a lengthy critique. In short, it said Hooke had performed all the same experiments, drawn different conclusions, and that Newton was outright wrong. In 1704, Newton finally published a full account of his theory of light in *Opticks*. Newton had already drafted a treatise covering much of this work by 1672, but he never published it. Being extremely sensitive to criticism, he waited – more than thirty years – until the death of his longtime rival Hooke. Newton's theory of light intertwined the color of light

---

[186] Specifically, he said, "a certain movement, or very rapid and lively action, which passes to our eyes through the medium of the air and other transparent bodies, just as the movement or resistance of the bodies encountered by a blind man passes to his hand by means of his stick."

[187] The exact timeline of Newton's experiments in optics is a bit confused. Nonetheless, Newton most likely began his work in optics around 1666, extended these initial investigations, and clarified his theory by 1669. Although by 1670 he had a fully elaborated theory, he only published his initial thoughts in a 1672 article.

with the idea that light was composed of particles, which Newton referred to as "corpuscles." Whereas his theory of colors was rooted in experimental results with prisms, light as a particle was merely a conception. Nonetheless, mostly due to Newton's prominence, his particle theory of light prevailed for more than a century after *Opticks*.

Newton thought the properties of light were best explained by considering light as a particle. For example, he felt his laws of motion could explain *refraction*, the property of light that causes it to bend when passing from one substance to another. For example, when passing from the air to water, light will bend (towards or into the water) at the surface. This bending of light is why an object, like a pencil, that is partially placed in water appears bent or "broken."

According to Newton, refraction occurs because the particles of light have their direction changed (causing the bend) due to the forces acting upon them at the interface of the two substances. For Newton and other proponents, a particle theory of light also explains *reflection*. If we imagine the particles to be like billiard balls, then reflection is nothing more than light particles colliding with a surface and "bouncing" off it, much like a billiard ball collides with the side of a pool table only to be bounced (or "reflected") back. In addition to this, Newton felt there were some properties a wave theory of light simply couldn't explain, such as *diffraction*.

Diffraction is the property of waves that allows them to bend around objects and spread through openings. In the case of sound waves, it's why a sound in one room can often be heard in another room by someone not directly in the path of the oncoming sound; the sound wave travels from one room, spreading out through the doorway, into the other room where it's then heard.[188]

Light doesn't appear to behave this way. After all, you can't see around corners, can you? The amount of diffraction depends on the *wavelength*

---

[188] A lower pitch sound – that is, a sound wave with a longer wavelength (lower frequency) – will diffract (or bend) around an object more than a higher pitch sound (a sound wave with a shorter wavelength and higher frequency). This means a lower pitch sound is more easily heard around an object that may be in front of the source of the sound than a higher pitch sound.

of the wave relative to the size of the opening it's passing through. If the wavelength is much smaller than the opening, the diffraction will be unnoticeable. However, the amount of diffraction will be considerable if the wavelength is comparable to the size of the opening, and even more so if the opening is smaller than the wavelength. The diffraction of light is usually unnoticeable because light waves have such short wavelengths – much smaller compared to the wavelengths of sound waves. Nonetheless, light will diffract if the opening is small enough. Francesco Grimaldi (1618–1663) discovered the diffraction of light around 1660.

Despite Newton's arguments, Hooke published a wave theory of light in the 1660s. In 1678, Christiaan Huygens developed a wave theory (publishing it in 1690 in his *Treatise on Light*) capable of explaining reflection and refraction. However, the beginning of the end for Newton's particle theory of light occurred in the early1800s when Thomas Young (1773–1829) provided the first clear demonstration of the wave nature of light, showing that it exhibits the property of *interference*. Under the right conditions, when light from two separate light sources combine, they will show regions that are brighter (*constructive interference*) and regions that are darker (*destructive interferences*) than either of the sources alone. Other waves, such as water and sound, also show this effect.

Imagine a stone dropped straight down into a still lake. It will cause ripples, or waves, to spread out from the point where it landed (the source). Now imagine doing this with two stones – perhaps one in each hand with your arms stretched out wide – at the same time. Clearly, waves will spread from two sources now. As the waves continue to spread, there will be some regions where the waves from the one source overlap with waves from the other source. Moreover, in these regions of overlap, the height of the wave (which is related to its *intensity*) may increase, while in other overlapped regions, the height of the wave may decrease. This illustrates constructive and destructive interference, respectively.

The interference nature of light, which was further confirmed by experiments and calculations performed by Augustin-Jean Fresnel (1788–1827) several years later, dealt quite a blow to particle theory enthusiasts. The final blow to the particle theory of light came with the determination of the speed of light in difference substances. According to the particle theory of light, the speed should be faster when light passes from one substance to another that is denser, as is the case when light moves from air to water. However, the wave theory of light predicted the opposite effect, that light should move slower in a substance of higher density, which was confirmed by 1850 with the work of Jean-Bernard-Léon Foucault (1819–1868) and Armand-Hippolyte-Louis Fizeau (1819–1896). Further developments on the nature of light in the nineteenth century led to the wave theory of light as generally accepted.

In 1864, James Clerk Maxwell published the last of a series of papers on electricity and magnetism. In this paper, he wrote the (now famous) four equations that form the basis of all of classical electricity and magnetism. This work and his two-volume book *Treatise on Electricity and Magnetism*, published in 1873, unified the forces of electricity and magnetism and brought the understanding of light to a whole new level. In his theory, Maxwell described light as an *electromagnetic wave*, and he was able to correctly calculate its speed:

> This [calculated speed] is so nearly that of light, that it seems we have strong reason to conclude that light itself (including radiant heat, and other radiations if any) is an electromagnetic disturbance in the form of waves ....

Although the speed of light had been measured,[189] light as an electromagnetic wave was a totally new concept. Unfortunately, Maxwell would not live long enough (dying in 1879) to see this aspect of his theory validated.

In 1887, Heinrich Hertz (1857–1894) confirmed Maxwell's theory by actually producing the electromagnetic waves it predicted. Further, Hertz and others showed that these waves displayed the anticipated properties of reflection, refraction, and diffraction. Thereafter, these results firmly established Maxwell's theory and the wave concept of light in essentially every physicist's mind. That is, until Einstein decided to shake up its very foundation in 1905.

---

[189] In 1862, Foucault measured the speed of light to be 299,796 km/s, which is in good agreement with the modern value of 299,792.458 km/s.

Photoelectric Effect

Ironically, in the same set of experiments he was conducting to investigate the wave nature of light described by Maxwell's theory, Hertz inadvertently discovered the photoelectric effect as well. The photoelectric effect was a bit of a burden on the venerated Maxwell theory of light. When light is shined on certain types of metals, electrons are emitted, or knocked free, from the surface (Figure 14.1). This phenomenon is the photoelectric effect, and several features of it were impossible to explain with Maxwell's theory:

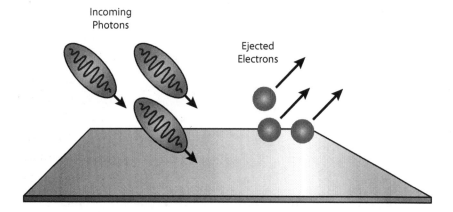

Figure 14.1. Incoming (incident) photons of energy, $h\nu$, strike the surface of a metal. As they do so, they knock electrons free from the metal's attractive pull.

- In order for electrons to be knocked free, the light striking the surface (the *incident light*) must be greater than a certain frequency, which is characteristic of the metal it's shining on. In other words, there's a cutoff frequency that the light striking the surface must be above; otherwise, there's no photoelectric effect. This is inconsistent with the wave theory of light, which predicts electrons should fly off at *any* frequency once the intensity is high enough.

- Once the light exceeds the cutoff frequency, electrons begin to fly off with a certain speed, or kinetic energy. While the number of electrons flying off is proportional to the intensity of the light, the kinetic energy, or speed with which they fly off with, is not. In fact, their kinetic energy is only affected by the frequency of the light, not its intensity.

- The electrons that fly off do so almost immediately after being hit by the light. That is, the effect is almost instantaneous. This occurs even when the intensity of the light is low.

For Einstein, the solution to all these issues was clear. As we discussed earlier, Einstein proposed that light consists of light quanta. In other words, light is made up of *particles* with energy hv, which today we call *photons*. Therefore, when light shines on a metal, a photon can collide with an electron in the metal and transfer its energy hv to it. This transfer of energy is almost instantaneous since the energy of the photon isn't spread out over a large area, but rather is concentrated, thus allowing the energy to be quickly passed to the electron. Now, when the photon bumps the electron it will send it flying off the metal, only if the energy of the photon is "enough." What do we mean by enough?

A given electron is held to the metal by the attractive interaction between it and the atom it's bound to; the positively charged nucleus of the atom holds the negatively charged electron and keeps it from simply flying off by itself.[190] Therefore, the photon needs to have enough energy to free the electron from this attractive pull. This amount of energy is the bare minimum that is needed – anything less won't be enough to free the electron. This explains the cutoff frequency mentioned earlier.

The photon needs to have a certain frequency, v – or a certain amount of energy hv (since h is a constant) – to free the electron from the grip the atom has on it. This minimum amount of energy is called the *work function*, W – it takes work for the electron to free itself – and it's inherent to the metal. Moreover, it explains why the photoelectric effect is seen only above a certain cutoff frequency, rather than at any frequency once the intensity is high enough according to wave theory.

According to the photon concept, a higher intensity would result in more photons of the same energy. Even though we have more photons, if

---

[190] This is a bit oversimplified. A given electron feels an attractive interaction resulting from all the atoms that it's proximal to, rather than simply the attractive interaction of a single atom.

they don't have the minimum amount of energy required (the work function amount, W) to free the electron from the atom, nothing will happen. What if the energy is greater than the work function? What happens to the leftover energy? After all, we know that it doesn't simply disappear since, as we learned in Part I, energy is conserved.

The leftover energy goes into the speed the electron flies off with. The more energy that is left over, the faster the electron speed will be as it flies off the metal. In fact, we need a bit more energy than simply that of the work function to get the electron to fly off. As mentioned, a higher intensity means more photons, and more photons with "enough" energy will cause more electrons to fly off the metal.

In other words, once the frequency cutoff is met, increasing the intensity will directly increase the number of electrons coming off the metal. However, it won't affect the speed with which they fly off; this is only affected by the frequency of the photon hitting the electron. Recall that the speed results from the "leftover energy" we talked about earlier. That amount and the work function amount, W, are equal to the total amount of energy the photon has, hv, which only depends on the frequency, v, of the photon. We can sum up all these results in a simple equation originally written by Einstein:

$$hv = K_{max} + W$$

where K_{max} is the maximum kinetic energy the electron can have. Clearly, this is just an equation for the conservation of energy, similar to those we talked about in Part I, with the only difference being that we are talking about energy conservation of microscopic particles – even they conserve energy.

Einstein's light quanta and energy conservation equation successfully explained the photoelectric effect and other effects that the wave theory of light and Maxwell's theory failed to. You might imagine that such a breakthrough would have been welcomed with open arms from the physics community. Not so, unfortunately.

A Lonely Endeavor

Einstein's photon concept met with tremendous resistance, and it took almost twenty years after its introduction to be fully accepted. The general sentiment of this resistance can be felt in this excerpt of the letter of recommendation written by Planck, Emil Warburg (1846–1931), Walther Nernst (1864–1941), and Heinrich Rubens (1865–1922) in 1913 for Einstein's acceptance into the Prussian Academy of Sciences: "That he may sometimes have missed the target in his speculations, as for example, in his hypothesis of light quanta, cannot really be held too much against him …."

The reason for the opposition was clear: Einstein had dared to call out the wave theory of light, which had seen much success with explaining physical phenomena, especially under the direction of Maxwell's theory. Therefore, tampering with something that had given so much in the way of explanation offended most physicists.

Planck also encountered resistance to his energy quanta, but nothing near what Einstein endured for his light quanta, or photon concept. Recall that Planck proposed that the quanta applied to the energy of his resonators in their interactions with light. From this one might reasonably assert that when matter and light interact, the result is that the energy of matter ends up always quantized. Planck never dared to touch the wave nature of light or Maxwell's theory by attempting to *quantize* the energy of light, as Einstein so boldly did. Planck wanted nothing to do with this – he was barely a proponent of his own theory – and he makes this very clear in a letter he wrote to Einstein in 1907:

I am not seeking the meaning of the quantum of action [light quanta] in the vacuum but rather in places where absorption and emission occur, and [I] assume that what happens in the vacuum [to light] is rigorously described by Maxwell's [theory].

And once again at a physics meeting in 1909:

I believe one should first try to move the whole difficulty of the quantum theory to the domain of the interaction between matter and radiation.

It's always been interesting to me that Planck was able to make the distinction of energy quanta as only applying to his resonators (in their interactions with light) and matter in general, but not to light itself. That is, for Planck light and matter could be in equilibrium, passing energy back and forth, but for matter this energy transfer (either absorption or emission) with light must be quantized, but for light it's not – for Planck, light was not quantized. Seems rather one-sided, doesn't it? Perhaps Planck didn't want to shake things up too much more than he already had.

Planck's refusal to infringe upon Maxwell's theory and the wave nature of light lowered the resistance to his version of quantum theory. Moreover, Planck had at his disposal top-notch experimental data, which not only aided him in developing his theory but also confirmed, without any doubts, his expression for the blackbody spectrum. In fact, it was a simple matter of taking Planck's expression and comparing it to the experimental data, which upon doing so, one undoubtedly found perfect agreement.

So, while one could doubt the way Planck arrived at this result – many physicists did – it was hard to imagine, even to the most skeptical of minds, that he wasn't doing at least something correct. Einstein had nothing of the sort. His main expression was the energy conservation equation we discussed. But unlike Planck and his expression for the blackbody spectrum, Einstein didn't have high-quality experimental data for comparison. He would have to wait more than a decade for such data.

When this data arrived in 1916, thanks to the work of Robert Millikan (1868–1953), he was less than gracious to Einstein's theory:

> Einstein's photoelectric equation ... appears in every case to predict exactly the observed results. ... Yet the [particle] theory by which Einstein arrived at his equation seems at present wholly untenable.

In a following paper, Millikan called Einstein's light quanta hypothesis "the bold, not to say the reckless, hypothesis of an electro-magnetic light [particle]."

Thus, even though his own experiments demonstrated the correctness of Einstein's photoelectric (energy conservation) equation, Millikan refused to believe light consists of particles called photons. Therefore, while Millikan didn't deny his results proved Einstein's equation correct, he was unwilling to believe that the underlying mechanism responsible for the equation had anything to do with Einstein's photons.

But this was pretty much the sentiment of the majority of physicists, and it's clearly illustrated in the citation for Einstein's 1921 Nobel Prize: "for his services to Theoretical Physics, and especially for his discovery of the law of the photoelectric effect."

Certainly, Einstein is being acknowledge for "the law of the photoelectric effect," in other words for his photoelectric equation, but not for the photon concept. This attitude would persist until 1923, when new experimental results would convert pretty much everyone to the existence of photons – more on this later.

Until then, Einstein was alone in this pursuit. Unwavering in his commitment, Einstein continued to explore the quantum nature of light, furthering the photon concept along the way, and quantum theory in general.

Reconsidering Planck

In his paper of 1905, Einstein distanced himself from Planck to develop his own version of quantum theory, namely that light consists of particles, or light quanta, which would later be called photons. Two of the biggest misconceptions about this paper are that it was exclusively about the photoelectric effect – it wasn't – and that Einstein was simply building upon Planck's work – also not true. In 1906, Einstein revisited his 1905 work in the context of Planck's, showing that in his derivation, Planck made implicit use of Einstein's light quanta concept. Reflecting on this, Einstein said:

At that time [in 1905 when I first proposed the light quanta hypothesis] it seemed to me that in a certain respect Planck's theory of radiation constituted a counterpart to my work. New considerations ... showed me, however, that the theoretical foundation on which Mr. Planck's radiation theory is based differs from the one that would emerge from Maxwell's theory and the theory of electrons, precisely because Planck's theory makes implicit use of the aforementioned hypothesis of light quanta.

Using the statistical mechanical methods he had developed in 1903, Einstein directly derived the entropy of Planck's system of resonators without requiring the use of Boltzmann's method, as Planck did. Instead, all that was needed for Einstein to complete the connection to Planck's version was to further postulate that the energy of a resonator could only have a value of energy that was an integral multiple of ε.

So, whereas Planck arrived at this conclusion through a comparison of his derivation for the entropy with that obtained with Boltzmann's method, Einstein simply needed only his own derivation and the light quanta hypothesis. Moreover, Einstein insisted that the energy of a resonator could only change in "jumps" through emission or absorption of a photon of energy $\varepsilon = h\nu$; if a resonator's energy goes down or up – as it does when a resonator emits or absorbs light – it must do so in increments of ε, no more, no less.

Thus, Einstein proposed a physical mechanism for the interaction between matter (resonator) and light, and he also established a relationship with his light quanta hypothesis. This more complete picture of the equilibrium between matter and light, which was absent from Planck's original version of his theory, endowed Planck's theory with real physical attributes.

The nature of light was a tremendous source of preoccupation for Einstein. In 1908, he wrote to a friend, "I am ceaselessly occupied with the question of the constitution of radiation …. This quantum question is so incredibly important and difficult that everyone should busy himself on it."

Indeed, Einstein was searching to understand the general properties of light. To this end, whereas in 1905 his focus was on light in the low-density or high-frequency regime, where Wien's distribution is perfectly valid, in 1909 he sought to understand light over its complete range of frequencies in accordance with Planck's distribution. What he found would change our view of light forever.

The Dual Nature of Light

In 1909, Einstein would once again call upon his statistical mechanical methods. In particular, he was concerned with the statistical mechanics of *fluctuations* incurred by a system, something he had considered originally in 1904. Fluctuations are the natural deviations a system sustains from the average value of a given property. For example, we often talk about the fluctuations in energy around the average value of the system. Physically, fluctuations result because the value of the property at a given instance comes from a specific distribution or range of values; the value isn't fixed forever. You're already familiar with the fact that a system of ideal gas atoms takes on a Maxwell distribution of velocities at equilibrium.

Moreover, in general, such a system will also explore a Boltzmann distribution with regard to the total energy. These are two well-known examples of physical distributions that describe a range of possible values the system can take on for the given properties, from which follows an average value along with a respective fluctuation (or variation) around this average as well. In 1904, Einstein had shown that such statistical reasoning applied to light as well. This was the beginning of several successes Einstein would have in applying statistical mechanics to light.

The 1904 version of Einstein's fluctuation method culminated in an equation that had already been derived by Gibbs but was apparently unknown to Einstein. In 1909, Einstein re-derived the equation, this time using an approach that was all his own. Applying his equation directly, he calculated the fluctuations in the energy for blackbody radiation (in a small frequency range). What he found was striking: the resulting expression for the energy fluctuations showed both a "wave term" and a "particle term."

In other words, with respect to its fluctuations in energy, light simultaneously behaves as both a wave and a particle (photon). He then went on to consider the fluctuations in momentum (radiation pressure). For this, he was unable to use his fluctuation equation. Rather, he arrived at his desired result by cleverly considering a "small" mirror enclosed in a box with light at a given temperature. He imagined the mirror moving only along one direction (imagine a mirror on a fixed track). As the mirror moves, it will suffer "collisions" with the light reflecting off it. These collisions will result in two effects. First, the constant bombardment of the mirror by the light will result in a drag, or frictional force, upon the mirror, which will slow it down, causing it to lose momentum. However, the collisions between the mirror and the light will be irregular. The irregular nature of these collisions means that rather than experiencing a constant force from the radiation, the mirror experiences a fluctuating force, which will actually cause the mirror to move (assuming it's small enough). The mirror's loss of momentum from the friction will be exactly compensated – on average – by the gain in momentum from the fluctuating force.

The scenario just described is analogous to our previous discussion of the Brownian particle (in Part III), which Einstein considered in detail in 1905. In the current scenario, the mirror plays the role of the "larger" Brownian particle being bombarded by the "smaller" particles of light, or photons. From his analysis of these effects from the light on the momentum of the mirror, Einstein was able to derive the momentum fluctuations of the light and found, as he had for the energy fluctuations, that there was both a wave term and a particle term.

Interestingly enough, in a second paper in 1909, Einstein once again considered the momentum fluctuations due to light. However, this time he considered not only the mirror and the light, but also added to the box an ideal gas. The result was exactly the same. Perhaps he needed to convince himself one more time of this remarkable result. From his efforts Einstein concluded:

> It is therefore my opinion that the next stage in the development of theoretical physics will bring us a theory of light that can be understood as a kind of fusion of the wave and [particle] theories of light.

Einstein was completely alone with his extremely bold sentiment. Planck was still struggling with the implications of his own theory, and the next big wave in quantum wouldn't arrive until 1913. The fusion of waves and particles under what would become *quantum mechanics* would finally arrive in 1925. But as we'll see, this version of "the fusion" would be one that dismayed Einstein to the point of severing all ties with an area of physics in which he had once led the way.

An inkling of Einstein's hesitations about quantum made a subtle appearance in 1917 as he finished three more impressive publications on the quantum theory of light. We'll discuss this in more detail, but we first must discuss the next big wave in quantum theory and how it changed our view of the atom forever.

CHAPTER 15

The Quantum Atom

Revisiting the Atom

Establishing the reality of atoms was a long and difficult process. Nonetheless, the theoretical work of Einstein in 1905 and the experimental work of Perrin that followed in 1909 (see Part III) finally secured the atom's existence once and for all. Indeed, it appeared the journey had come to an end, but actually it was just beginning.

The almighty atom had been deemed the fundamental particle of matter, the very building blocks themselves. Consequently, the atom was presumed to be indivisible (it's what "atom" means, after all). However, it began to appear that the atom consisted of fundamental pieces of its own, endowing it with an internal structure, which, under certain conditions, was quite "breakable" after all.

A glimpse into the atom's interior occurred in the 1830s with the work of Michael Faraday (1791–1867), whose experiments in electrochemistry led him to reflect on the composition of atoms:

> if we adopt the atomic theory or phraseology, then the atoms of bodies which are equivalents to each other in their ordinary chemical action, have equal quantities of electricity naturally associated with them.

Nonetheless, in the 1830s it was hard enough to be pro-atom, let alone to consider atoms themselves as having internal structure that gave rise to "electricity," as Faraday was speculating:

> But I must confess I am jealous of the term *atom*; for though it is very easy to talk of atoms, it is very difficult to form a clear idea of their nature, especially when compound bodies are under consideration.

Thus, he resorted to the "more acceptable" concept of electricity as an *imponderable fluid*, putting his mind at ease about the possibility of electrically charged particles associated with atoms. Decades went by until the next peek inside the atom occurred.

In 1895, Wilhelm Röntgen (1845–1923) discovered a new form of radiation he called "X-rays" (the "X" denoting the unknown nature of the radiation). Although he was unable to determine the physical mechanism behind their production, he found that X-rays did have the remarkable ability to penetrate almost anything they hit, including body parts, which he found – as we know today – allows for a "snapshot" of one's bones.

This discovery created quite a stir. In 1896, following up on a hunch that certain uranium compounds might also emit X-rays, Henri Becquerel (1852–1908) discovered that the uranium salt, potassium uranyl sulfate, did in fact spontaneously emit X-rays. The timing couldn't have been better for Marie Curie (1867–1934), who was looking for a topic for her doctoral thesis. Her work with uranium salts revealed that the amount of radioactivity was directly proportional to the amount of elemental uranium in the sample.

Moreover, she hypothesized that the X-ray radiation was coming directly from the uranium atom itself, pointing to an internal structure within the atom. It was still speculation, but a major unveiling of the inner pieces of the atom was just around the corner.

Going Subatomic

Joseph John ("J. J.") Thomson (1856–1940) had been conducting experiments with the cathode ray tube, which was a very popular apparatus at the time. It's a glass tube that has been evacuated of most of the air or other gas inside with two metal terminals at each end. By connecting a high-voltage source to each terminal of the tube, a flow of electricity starting from one terminal (cathode) to the other (anode) is produced. If the gas pressure inside is low enough, the gas inside will glow; cathode ray tubes are the forerunner to the modern-day neon sign and fluorescent light bulb. Lowering the gas pressure even further will cause the glowing to disappear, but the flow of electricity will continue. While this stream itself is invisible, its presence is apparent by the glow that results at the other end of the tube as it strikes the glass. Moreover, an object place inside the tube in front of the cathode will cast a shadow on the glowing glass. The nature of this stream of electricity had been of interest to many for some time.

In 1897, Thomson showed that this stream consisted of negatively charged particles much smaller than the smallest atom (hydrogen), which later became known as electrons. The electrons break free from the atoms constituting the metal of the cathode itself. In fact, modern versions of the cathode ray tube heat the cathode to such a high temperature that the electrons are actually set free by being "boiled off" (thermionic emission). The gas inside the tube serves to conduct these electrons much the same way that an electrical wire conducts electrons in everyday appliances like a coffee maker, TV, or stove. Conduction occurs by electrons being passed from one atom to another.

A key finding in Thomson's work was that regardless of the metal used for the cathode or the gas contained within the tube, the charge-to-mass ratio of the electron was always the same. This finding convinced him that electrons were common to all atoms. Not only were they common to all atoms, they were identical for all atoms regardless of the atom type, or element. Thomson had found the first subatomic particle. Reflecting on it, Thomson said:

At first there were very few who believed in the existence of these bodies smaller than atoms [i.e., electrons]. I was even told long afterwards by a distinguished physicist who had been present at my lecture at the Royal Institute that he thought I had been "pulling their legs." I was not surprised at this, as I had myself come to this explanation of my experiments with great reluctance, and it was only after I was convinced that the experiments left no escape from it that I published my belief in the existence of bodies smaller than atoms.

Thomson didn't rest on his laurels, although he was well within his rights to do so. Being convinced that at least part of the internal structure of all atoms consisted of electrons, he set out to construct a model of the atom, like so many others before him (including Dalton). However, unlike all the others, Thomson was the first to include electrons in his model. As Thomson knew, a good model must, at least, account for the neutral charge of the atom and the mass.

Accounting for the neutral charge is relatively straightforward: electrons are negatively charged, therefore there must be a surrounding positive charge that will cancel out the total negative charge of all the electrons. A two-dimensional version of Thomson's model would resemble a chocolate chip cookie, where the electrons are the chips and the positive charge is the rest. He attributed the majority of the mass of the atom to the electrons (that's some pretty heavy chocolate chips).

Thomson debuted his model in 1903. Although a decent start, it was plagued with problems. Nonetheless, it stuck around for quite some time, until new experimental results would begin to crumble Thomson's model in 1909.

In 1895, Ernest Rutherford (1871–1937) arrived in Cambridge on a research fellowship to work with Thomson. For Rutherford, it was an invigorating and opportune time to be a young experimental physicist. It was a mere three months after his arrival that Röntgen published his discovery on X-rays; then came Becquerel's initial paper on radioactivity three months later; and Thomson's announcement of the electron a year thereafter. In 1897, Rutherford pursued his own research on radioactivity, and in 1898 announced his discovery of two new forms of radiation: α-particles and β-particles.

These discoveries are all the more amazing given the extremely primitive tools (by today's standards) at his disposal. In 1898, he took a position as professor of physics at McGill University in Montreal. Indeed, Rutherford was a superstar: in 1903 he was elected to the Royal Society; in 1905 they honored him with the Rumford Medal; and in 1908, at only thirty-seven years of age, he won the Nobel Prize in chemistry "for his investigations into the disintegration of the elements, and the chemistry of radioactive substances." However, his greatest success was still yet to come.

In 1907, he took a professorship at the University of Manchester. Under Rutherford's guidance, Hans Geiger (1882–1945) and a twenty-year-old undergraduate student, Ernest Marsden (1889–1970), studied the scattering of α-particles by thin metal foils in 1909. They observed that occasionally the α-particles would be scattered by more than 90°. The result was amazing, Rutherford himself said, "It was quite the most incredible event that ever happened to me in my life. It was almost as incredible as if you had fired a 15-inch shell at a piece of tissue paper and it came back and hit you." He goes on to say, "I did not believe they [the α-particles] would be [scattered], since we knew that the α-particle was a very fast massive particle, with a great deal of energy."

So, as an α-particle passes through a metal foil, an atom in it somehow causes a significant amount of scattering. In fact, a scattering angle of greater than 90° is only possible if the thing doing the scattering (metal atom) is more massive than the thing being scattered (α-particles).

In Thomson's (chocolate chip cookie) model of the atom, the electrons reside in a vast, spread-out body of positive charge. However, an α-particle weighs about 8000 times more than an electron, so there's no way that an electron of a metal atom could be responsible for the scattering. That leaves the positive charge, which must be much less spread out and more concentrated than Thomson was imagining. A year and a half later, in December 1910, Rutherford solved the mystery. Rather than being spread out inside an atom, the positive charge was a very compact, central body that electrons orbit around. Further, this centralized location is where most of the mass of the atom resides – contrary to Thomson's assertion that the electrons accounted for the majority of an atom's mass.

Rutherford went further and derived a mathematical formula for the scattering process, which was later verified by a series of experiments performed, once again, by Geiger and Marsden. Rutherford's formula is even more impressive given that he was an experimentalist, who didn't much care for or have expertise in theoretical physics; no theoretical physicist had responded to the challenge. Recall that Geiger and Marsden *occasionally* saw the back scattering of an α-particle. In other words, most of the α-particles passed right on through with hardly a deflection in their path. This means that there's quite a bit of space between the central positive charge, or *nucleus*, and the surrounding electrons swarming around it.

Rutherford's model of the atom didn't catch on immediately. Even Rutherford himself appeared not to take it too seriously. During the first Solvay Conference – a major international science conference held in 1911 – he mentioned nothing of his new atomic model. A year later, he completed a 670-page book *Radioactive Substances and Their Radiation*, yet devoted merely three pages to the α-particle results. Perhaps Rutherford didn't see these discoveries as groundbreaking, but there was one man who realized their importance and pushed Rutherford's atomic model to the next level.

In March of 1912, Niels Bohr (1885–1962) arrived in Manchester to begin working with Rutherford. Previously, he had worked with Thomson in Cambridge. Unfortunately, their relationship had been strained from the start and never really flourished as Bohr had hoped. Writing to his brother Harald, Bohr said, "Thomson has so far not been as easy to deal with as I thought the first day."

Perhaps Bohr's initial encounter with Thomson was to blame, where upon entering Thomson's office, Bohr proclaimed: "This is wrong."

He was referring to something in a book Thomson had written. Of course, Bohr never meant to be insulting; rather, he was simply trying to engage in a scientific discussion with his limited command of English (at the time). The situation was undoubtedly exacerbated by Thomson's inability to tolerate criticism. Later in life, Bohr reflected on his time with Thomson in these words: "The whole thing was very interesting in Cambridge but it was absolutely useless." But his circumstances significantly improved under Rutherford:

Rutherford is a man you can rely on; he comes in regularly and enquires how things are going and talks about the smallest details – Rutherford is such an outstanding man and really interested in the work of all the people around him.

Bohr's time with Rutherford was brief, lasting only about four months. During this time Rutherford was cautiously encouraging of Bohr's efforts, although quite preoccupied with his own endeavors of book writing and new research interests. In fact, Bohr's main scientific influence wasn't Rutherford. Instead, the new physics Bohr learned came from two other researchers under Rutherford: Georg von Hevesy (1885–1966) and Charles Galton Darwin (1887–1962). Regardless, Rutherford's model of the atom would inspire and provide Bohr with a tremendous stepping-stone to his own work on atoms, by which he would ultimately be forever known. Bohr wrote to Harald from Manchester, "Perhaps I have found out a little about the structure of atoms." What an understatement this turned out to be.

Bohr's Atom

Rutherford's model was a big step towards understanding the atom. From the experimental data he and his researchers collected, the most plausible picture of the atom was one where a very compact nucleus resides at the center, surrounded by electrons flying all around it. More precisely, we can imagine the electrons orbiting the central nucleus similar to the way the planets orbit the Sun. Unfortunately, this version of an atom is unstable.

According to the known physics at the time (classical physics), electrons moving in such a fashion would emit light, which is a loss of energy for an electron. This energy loss shows up in the form of a lower potential energy for the electron, which means it moves closer to the nucleus.

To understand the potential energy a negatively charged electron "feels" for the positively charged nucleus, imagine a rubber band that has been fixed to a wall at one end, while we begin to stretch the other end outward. As we keep stretching, there will be a point where we'll feel the tension in the rubber band resisting being stretched further by pulling inward against us. At this point, the potential energy is very high, but if we stop stretching and begin to let the rubber band move inward, the resisting tension goes down, along with the potential energy.

We can imagine the potential energy between an electron and the nucleus as resulting from an invisible "rubber band" with the electron connected to one end, and the nucleus connected to the other while fixed at the center of the atom. What prevents the electron from being completely pulled into the nucleus is that the inward pulling of the "rubber band" and the outward stretching of the centrifugal force equal each other.

The real problem is that the electron continues to emit light, thus losing energy, and moves closer and closer to the nucleus, until it finally collides with the nucleus and destroys the atom. Such was the fate of Rutherford's (classical physics) version of the atom. Bohr wasn't bothered at all.

The failures of classical physics were already very familiar to Bohr from his PhD work. So seeing it fail in the realm of atoms was not much of a surprise to him: "This seems to be nothing else than what was to be expected as it seems rigorously proved that the [classical physics] cannot explain the facts in problems dealing with single atoms."

How did Bohr reconcile the seemingly unquestionable Rutherford atom with the instability predicted by classical mechanics? By introducing a new hypothesis, "there will be given no attempt of a mechanical [classical physics] foundation …."

Bohr hypothesized that the *binding energy* of the electron – the energy required to remove an electron from the very atom that holds it – can only come from a set of discrete, rather than continuous, values. In other words, just as it was for the energies of Planck's resonators, the binding energies for the electrons in Bohr's atom also take on a quantized form:

$$E_b = Cnh\omega$$

where E_b is the binding energy, C is a constant, ω is the rotational frequency of the electron in its orbit, which is simply the velocity divided by the total distance it travels around the orbit (the orbit was assumed to be circular, so this distance is simply the circumference), and $n = 1, 2, 3$, and so on.

What's striking about this equation is its likeness to Planck's equation for the energy of a resonator:

$$E_{\text{resonator}} = mh\nu$$

Recall that ν in Planck's equation is the vibrational frequency of the resonator (once again, ω in Bohr's equation is the rotational frequency of the electron), and $m = 0, 1, 2, 3$, etc. Thus, Bohr makes a formal analogy to Planck's energy quantum, and in doing so gives it a real physical meaning. Later in life, Bohr remarked, "It was in the air to try to use Planck's ideas in connection with such things."

In addition to quantizing the binding energy, Bohr also obtained results showing that an electron's distance from the nucleus, or the location of its orbit, is also quantized (as is its angular momentum, or orbital momentum).

The physical picture embodied in Bohr's atom is one where the electrons surrounding the nucleus reside in discrete orbitals with discrete energies. As before, by "discrete" we mean quantized, and for Bohr's atom this applies to both the orbitals and the energies, whereas for Planck's resonators it was simply the energy that was quantized. The quantization is directly due to the quantum number, n, and larger n corresponds to an orbital farther away from the nucleus with a greater binding energy.

Although Bohr's quantum number, n, bears a relationship to m in Planck's equation, its role is more substantial. The quantum number is describing the actual *quantum state* of the electron, and it's only in those particular quantum states that an electron is stable, according to Bohr's hypothesis, and therefore won't suffer the demise of spiraling towards the nucleus. Notice that, unlike Planck's equation, where m can be zero, in Bohr's equation n can't be zero, since this would correspond to the quantum state where the electron has fallen into the nucleus, and once again we have the atom's demise.

At the time of Bohr's theory, over fifty years had passed since the work of Kirchhoff and Bunsen revealed that the spectra of atoms emit a unique fingerprint consisting of discrete spectral lines, which also represent the exact same frequencies that the atom will absorb at. While the experimental side of spectroscopy continued to make significant progress over those years, for theory it was a different story.

Thomson's discovery of the electron instigated speculation on their role in an atom's spectrum, but no real progress had been made. It was beginning to appear that a theory would never be found. This sentiment is well expressed by the physicist Arthur Schuster (1851–1934), who in 1882 said:

> It is the ambitious object of spectroscopy to study the vibrations of atoms and molecules in order to obtain what information we can about the nature of forces which bind them together .... But we must not too soon expect the discovery of any grand and very general law, for the constitution of what we call a molecule is no doubt a very complicated one, and the difficulty of the problem is so great that were it not for the primary importance of the result which we may finally hope to obtain, all but the most sanguine might well be discouraged to engage in an inquiry which, even after many years of work, may turn out to have been fruitless.

Bohr had provided a draft of his initial ideas on atoms (our previous discussion summarizes the most significant portions) to Rutherford in July of 1912, but never mentioned a single word about atomic spectra. Almost a year would go by until Bohr would seriously consider atomic spectra in the context of his theory. His interest was sparked by an engaging conversation he had, in early February of 1913, with H. M. Hansen (1886–1956).

Hansen had worked on spectroscopy in Göttingen. He inquired about Bohr's efforts on using his theory to predict spectra. Bohr mentioned that he hadn't really considered it, as progress there seemed unlikely. Hansen insisted that Bohr reconsider and pointed him in the direction of an intriguing spectral formula known as the *Balmer Formula*.

Jumping Electrons: Spectra

In 1849, Johann Balmer (1825–1898) received his PhD in mathematics from the University of Basel, Switzerland. He would remain in Basel his whole life, teaching at a girls' school and lecturing at the university. Being an enthusiast of numerology, Balmer believed that pretty much everything (like the number of sheep in a flock, the number of steps on an Egyptian pyramid, etc.) in life had some sort of special relationship to numbers and formulas. Although a mathematician by training, Balmer made no significant contributions to that field but helped further physics with his spectral formula for the hydrogen atom.

His accomplishment is nothing short of remarkable. At the time, Balmer knew of only four spectral frequencies for hydrogen that had been experimentally determined by Anders Ångström (1814–1874). Using only these four data points, Balmer, at the age of sixty, constructed a formula that predicted the complete frequency spectrum of the hydrogen atom (and also correctly predicted a lower and an upper bound over the range of frequencies). Shortly thereafter, Balmer learned that not only did his formula account for the original four frequencies, it also correctly predicted twelve other known frequencies.

In 1885, he published two papers on his efforts, which immortalized his name. Many more spectral frequencies have been determined over the years, and Balmer's formula remains intact. In 1890, Johannes Rydberg (1854–1919) found that Balmer's formula is actually a special case of a more general formula (that we now call the *Rydberg Formula*):

$$v = cR\left(\frac{1}{b^2} - \frac{1}{a^2}\right)$$

where v is the frequency, c is the speed of light, and R is Rydberg's constant. Balmer's frequencies for the hydrogen atom are obtained by setting $b = 2$ and letting $a = 3, 4, 5, 6$, and so on. Although this formula correctly predicted the spectral lines seen in the hydrogen atom, nobody knew why it worked – it just did. For almost thirty more years, the atom would keep this secret.

Following Hansen's suggestion, Bohr looked up Balmer's formula. He probably found it in the general form written by Rydberg (just like we've written above). The formula was widely known, and most likely Bohr had seen it as a student, only to have it slip his mind later. We have already discussed that Bohr proposed that the binding energy of an electron in an atom is quantized, much the same way that Planck quantized the energy of his resonators. Using this hypothesis, with some very straightforward classical physics, Bohr was able to arrive at an overall expression for the difference in binding energies, ΔE, between two quantum states, with quantum numbers n_1 and n_2:

$$\Delta E = K\left(\frac{1}{n_1^2} - \frac{1}{n_2^2}\right)$$

Upon seeing the Balmer Formula again, he must have immediately known how to coax out the physics buried deep within the formula all those years. The next hypothesis that Bohr advanced was that the energy difference between electron quantum states is equal to the energy of Einstein's light quanta, hv:

$$hv = K\left(\frac{1}{n_1^2} - \frac{1}{n_2^2}\right)$$

and thus he derived the Balmer Formula, where $R = K/hc$. In fact, Bohr was able to calculate the Rydberg constant and found excellent agreement with the experimental value. Moreover, the a and b of Balmer's formula are identified with the quantum states n_1 and n_2, rather than being simply integers with no physical significance. This provided tremendous support for Bohr's new version of his theory. Let's take a closer look at the physical insight behind Bohr's derivation.

By the time Bohr saw Balmer's theory, he already had what he needed to derive the binding energy difference, ΔE. However, seeing Balmer's formula is what most likely drove him to do so. After this, only one more step remained to complete the derivation. Bohr had to equate the binding energy difference to the energy of a single photon as given by Einstein's light quanta, arriving at

$$\Delta E = hv$$

which is known as the *Bohr Frequency Rule*.

The mathematics of this step is trivial (one simply equates ΔE to $h\nu$), but Bohr's physical interpretation is astounding. According to Bohr, an electron can change quantum states by "jumping" from one to another. If an electron goes from a higher to lower energy quantum state, it will emit a single photon, which will show up in the atom's frequency spectrum. In other words, an atom's frequency spectrum is due to all the "excited" electrons "jumping" from higher to lower energy quantum states, whereby they emit photons. Absorbing a single photon causes the electron to go from a lower to higher energy quantum state. Realize that given two quantum states, the energy, or frequency, of the photon emitted when the electron "jumps down" is the same energy, or frequency, of the photon absorbed when the electron "jumps up." Therefore, more than fifty years after Kirchhoff, Bohr's model finally explained the conclusion he made from his experiments about a substance: like an atom, it will emit and absorb at the same frequency.

Another success of Bohr's theory was the correct prediction of the *Pickering Series*. In 1896, Edward Charles Pickering (1846–1919) discovered a series of spectral lines in starlight that he attributed to hydrogen. The problem was that the Balmer Formula didn't predict these lines. Around the time Bohr was putting together his theory, the Pickering Series found renewed interest in the work of Alfred Fowler (1868–1940), who also found the spectral lines from laboratory experiments. Using his theory, Bohr was able to correctly account for the series of lines by attributing them to the frequency spectrum of singly ionized helium.

In the singly ionized state, helium is very "hydrogen-like" in the sense that (like hydrogen) it has only a single electron in the atom. It differs from hydrogen in the nucleus, where it contains two protons (to hydrogen's single proton) that give it a total positive charge of two (for hydrogen it's one), along with two neutrons (whereas hydrogen has none). This was another major success of Bohr's theory. Einstein's initial response to Bohr's theory was faint praise, but when he heard that it could correctly predict the Pickering Series his tone shifted: "This is an enormous achievement. The theory of Bohr must then be right."

There were other successes. Using his theory, Bohr correctly concluded that X-rays arise from an innermost electron jumping down to an orbital, or lower energy quantum state, left vacant by an electron previously knocked out of the atom. Moreover, Bohr initiated the first step towards quantum chemistry by realizing that the chemical properties of atoms arise from only their outermost, or *valence electrons.*

Bohr's paper on hydrogen was published in 1913, and two others quickly followed. The reception of his theory was mixed. Perhaps one of the most unsettling questions posed to his theory was by Rutherford, who wondered how an electron in one quantum state "decides" which quantum state it's going to jump into next. The point is that typical ideas of cause and effect (causality) seem to be missing from the theory. Indeed, this inherent issue with quantum mechanics would show up again when Einstein would revisit the interaction of light and matter.

Atoms and Light

By 1911, Einstein had already hypothesized that light consists of particles he called light quanta (later called photons). Moreover, he had shown that light has an inherent quality, whereby it exhibits both wave and particle properties. Although he had seen further than anyone into the mysterious nature of light, it continued to perplex him: "I do not ask anymore whether these [light] quanta really exist. Nor do I attempt any longer to construct them, since I now know my brain is incapable of advancing in that direction."

However, Einstein would content himself with light's strange behavior in order to focus on general relativity until November 1915, returning to light once again in July 1916. The end result was a deeper understanding of the interaction between light and matter, which culminated in three papers, two in 1916 and the most prominent one in 1917.

As we have seen, Planck pioneered the quantum theory of light and matter. In his model, matter took on the intentionally ambiguous[191] form of "resonators" – nothing more than an oscillating charge of small mass. A resonator's interaction with light was "mostly" classical, in that its explicit

---

[191] Planck intentionally kept the description of his resonator vague and used it to his advantage throughout his work on the quantum theory of light and matter.

interactions were between it and the classical electric field of light. The quantum portion of the theory was in the energy of the resonators, and its appearance in the theory was rather startling and without a mechanistic explanation. Bohr had brought quantum theory to the inside of the atom by quantizing the electronic orbits. He also provided a quantum theory of light and matter (at the atomic scale), where an electron jumping between orbits results in either the emission or absorption of light. Einstein's quantum theory of light and matter would extend beyond both of their theories while bringing together their best features.

Einstein was motivated by several factors. Undoubtedly, quanta and the "wave-particle duality" of light continued to weigh heavily on his mind. Bohr had provided a mechanism for the interaction of light and matter at the atomic scale, much richer in detail than Planck had with his resonator and light model, and Einstein wanted to explore the consequences further. Finally – and this point should be emphasized – it had been sixteen years since Planck derived his radiation law, and yet a completely quantum mechanical derivation of it was still nonexistent. Einstein says:

> [Planck's] derivation was of unparalleled boldness, but found brilliant confirmation. ... However, it remained unsatisfactory that the [classical mechanical] analysis, which led to [Planck's Radiation Law], is incompatible with quantum theory, and it is not surprising that Planck himself and all theoreticians who work on this topic incessantly tried to modify the theory such as to base it on noncontradictory foundations.

Indeed, despite its incredible success against ever-increasing experimental data, Planck's theory was still tarnished by its "mostly classical" derivation. In fact, to be honest, the rigorously derived parts of the theory were all classical in nature; the quantum component (centered on the energy of the oscillators) wasn't derived at all but was complete speculation.

In 1905, Einstein arrived at light quanta by comparing an ideal gas to light. At that time, he had considered them as two separate systems, each at thermal equilibrium contained in their respective boxes. In 1916–1917, he would consider them again, but this time as a "mixture" at thermal equilibrium in a single box. This time he explored what frequency distribution of light (frequency spectrum or radiation law) was needed to maintain this system of matter and light in thermal equilibrium.

Bohr's atom gave rise to the notion that an atom has electronic energy levels that electrons can jump between as they absorb or emit a photon. Thus, if we imagine the simplest atom – hydrogen, with its single electron – the quantum state of the entire atom is described by the electronic energy level occupied by the electron.

Einstein considered a collection of such atoms consisting of only two electronic energy levels. By now, Einstein was a master at using statistical mechanics to probe physical problems, and this time would be no exception. He asserted that the probability of finding the atoms of the system in either one of these two quantum states was given by the Boltzmann distribution. Moreover, he assumed that only three dynamical processes governed the transitions between the two energy states, each occurring with a certain transition probability. These three processes he named spontaneous emission, stimulated emission, and absorption.

Spontaneous emission occurs when an electron jumps from the higher to the lower energy state of the atom, emitting a photon in the process. The electron isn't actually influenced to do this by light but does it naturally. As with other spontaneous processes (recall from Part II), spontaneous emission is an irreversible process occurring naturally (without the slightest input of work) and resulting in an increase in entropy. On the other hand, stimulated emission occurs as a result of interacting with the light. Specifically, a photon "bumps" the electron on the way by, causing it to jump into the lower energy state, once again emitting a photon (in addition to the one that passed by in the first place). Finally, absorption occurs when an electron absorbs a photon and jumps up to the higher energy state.

Using these three processes as the basis for the interaction of light and matter in thermal equilibrium, Einstein arrived at the desired frequency distribution and found it to be given by none other than Planck's Radiation Law, the law describing the equilibrium spectrum for light in thermal equilibrium with matter. Of the three processes, stimulated emission hadn't been described before Einstein's 1916 work; the other two were already contained in Bohr's model.

It turns out that stimulated emission is crucial in obtaining the correct form of the radiation law; without it one obtains Wien's Radiation Law. Evidently, stimulated emission is important in correctly obtaining the low-frequency contributions to the radiation law. Another gem of Einstein's theory was that Bohr's frequency rule was obtained as a natural consequence. However, Einstein wasn't finished yet.

As discussed before (see Part II), a system of ideal gas atoms at thermal equilibrium will have a Maxwell distribution for the velocities. In 1917, Einstein asserted the same was true for thermal equilibrium of his mixture of atoms and light, and he set out to find the frequency distribution that made this possible. He used the same approach as in 1909 when studying momentum fluctuations of light.

Recall that he considered a "small" mirror, moving in only one direction, and light in thermal equilibrium. For his current study, an atom from the mixture plays the role of the small mirror, and the resulting fluctuation equation is the same. Using this as his starting point, he was able to show that Planck's Radiation Law is the correct form of the frequency distribution needed to maintain the Maxwell distribution for his mixture at thermal equilibrium. The result is particularly interesting because Einstein arrived at it by only considering the interactions between the atoms and light.

In other words, the collisions between the atoms themselves played no role in his calculation. Therefore, while a collection of ideal gas atoms alone at thermal equilibrium attain a Maxwell distribution for the velocities, so do the gas atoms of Einstein's mixture interacting with light (according to Einstein's three processes) at thermal equilibrium.

However, one of the most important consequences of Einstein's theory centers on the momentum of a photon. In Einstein's 1905 work about light quanta, the focus was on the energy of a photon of light, while its momentum was of no consequence. In 1909 this changed, and Einstein showed that the momentum fluctuations due to light involved both particle and wave terms. This was an amazing result that positioned Einstein to be able to write down both the energy and the momentum of a photon. He didn't do this at that time, and as to why he didn't, we'll never know.

It wasn't until his works of 1916–1917 that he completed the picture of the photon by endowing it with both energy and momentum. The momentum played a key role in Einstein's aforementioned result involving the Maxwell distribution. Einstein notes that, "The most important result [of the present work] seems to me, however, to be the one about the momentum transferred to the molecule in spontaneous or induced radiation processes."

According to Einstein, regardless of if the atom absorbs or emits a photon, the amount of momentum transferred is hv/c, the photon energy divided by the speed of light.[192] For absorption to occur, one imagines an incoming photon traveling in a specific direction and "bumping" into the atom. It's in this direction that the momentum is transferred to the atom.

For stimulated emission, the atom is once again "bumped" by an incoming photon, but this time it causes the atom to emit a photon of its own. As before, the direction of the momentum transfer is determined by the incoming photon, but this time it will be in the opposite direction. However, in spontaneous emission there isn't an incoming photon involved – the atom simply emits a photon at will.

So in what direction is the momentum transfer? According to Einstein, the direction is determined only by "chance." There's simply no way of knowing for sure as we did for the other two processes. Indeed, Einstein had hit upon the uncertainty that is inherent in what would later be quantum mechanics. This work would serve as a turning point in two regards. First, it would establish the physical reality of light quanta for Einstein, once and for all. Writing to a friend shortly after this work, Einstein says, "I do not doubt anymore the <u>reality</u> of radiation [light] quanta, although I still stand quite alone in this conviction."

---

[192] The momentum of light (as a wave) can be obtained from Maxwell's equations, which also yield the result to be the energy divided by the speed of light. Finally, the same result is obtained from Einstein's energy-momentum relation from special relativity, once the rest mass of a photon is set to zero.

Einstein remained alone in this conviction until 1923, when the experimental work of Arthur Compton (1892–1962) showed "very convincingly that a radiation [light] quantum carries with it directed momentum as well as energy."

Finally, his 1917 work would be the beginning of Einstein's departure from what would later be (the post-quantum theory of) quantum mechanics. Einstein had commented that the chance, or probabilistic, nature of spontaneous emission was a flaw of the theory, although he maintained confidence in the approach he had used. Once again writing to his friend, Einstein says, "I feel that the real joke that the eternal inventor of enigmas has presented us with has absolutely not been understood as yet."

If in 1917 Einstein truly saw the probabilistic nature as a defect in his theory, later he would be less forgiving. His last contribution to quantum theory (and considered by many to be his last significant scientific contribution) would occur in 1925, and thereafter he would turn his back forever on quantum mechanics, arguing that its probabilistic nature was a fundamental flaw.

CHAPTER 16

Quantum Mechanics

Nature's Lottery

Quantum theory officially started with Planck in 1900. In his revolutionary work he showed that an atom or molecule ("resonator" in his words) is only allowed discrete values for its energy. In other words, energy comes in "chunks" called quanta, just as matter comes in "chunks" called atoms. This behavior of energy is only noticeable when objects are very small, like atoms, molecules, and photons.

Indeed, it isn't something we notice in the objects of our everyday lives. For example, our cars aren't forced to move in "quantum jumps" like the electrons of Bohr's atom. Neither is any of the objects that we are exposed to on a daily basis – nor are we, for that matter. Nonetheless, the atoms that make up these objects are forced to move according to the very strange laws of quantum mechanics.

Einstein's 1917 work on the interaction of light and atoms touched on this strange behavior with the realization that it's impossible to know with any certainty the direction of the momentum transferred to an atom after a photon is given off during spontaneous emission. It's left up to "chance," or rather the (then unknown) laws of probability described by quantum mechanics; "chance" is inherent to quantum mechanics.

The use of probabilities to solve physical problems is already famil-iar to us. For example, we know a system of ideal gas atoms at thermal equilibrium will have a Maxwell distribution in velocities (while the total energy will be given by a Boltzmann distribution). In this way, we use probabilities to simplify calculations that otherwise would be very difficult or impossible. To be sure, it isn't that the velocities of the atoms of an ideal gas are inherently unknown, thereby forcing an uncertain description in terms of a probability distribution (e.g., the Maxwell distribution). On the contrary, probability is simply being used as a tool, rather than describing the underlying physical nature.

But in 1917, Einstein had uncovered the startling reality that nature does leave some things inherently unknown, and the probability associated with these things is more than a convenient calculation tool. It's a physical reality.

Together at Last: Photons and Planck's Radiation Law

As mentioned before, Planck's Radiation Law was derived mostly using classical mechanics and ending with his energy quanta hypothesis. However, a true quantum theory would have been devoid of the artifacts of classical mechanics altogether. In part this motivated Einstein's 1916–1917 works on the interaction of light and matter. Indeed, Einstein was able to arrive at a "much more" quantum derivation of Planck's Radiation Law. In the end, he too fell a bit short, forced to make assumptions (such as the Bohr Frequency Rule, Wien's Displacement Law) that prevented the derivation from being fully quantum. Actually, Einstein was poised to provide a full quantum derivation of Planck's Radiation Law in 1905.

At that time, Einstein imagined light at thermal equilibrium in a box, calculated the entropy, and found that the resulting expression was of the exact same form as for a collection of ideal gas particles (also at thermal equilibrium in a box). This intriguing similarity led Einstein to the conclusion that at low density, light behaves like particles (photons) rather than waves.

This new description of light provided a means of probing deeper into its physical properties, and for almost twenty years Einstein was alone in exploring the prospects. Nonetheless, one exceptional opportunity managed to slip by. Recall that Planck, in similar fashion to Einstein, derived an expression for the entropy of one of his resonators. He then followed it by a very crucial calculation: the number of microstates, W, for a collection of resonators. He did this by appealing to Boltzmann's method. Since Boltzmann's method requires one to first imagine the energy of the particles (in Planck's case, the resonators) as occurring in "chunks," the calculation became essential in shaping Planck's physical concept of the energy quanta.

For Planck, this extra step made all the difference since it led him to a novel physical interpretation, where before he had none. Whereas Boltzmann's method led Planck to energy quanta, Einstein's light quanta (he already had the "chunks" to begin with)[193] didn't lead him to Boltzmann's method. He never determined the number of microstates for his system of photons with this approach.[194] The benefit of doing so would have been nothing short of finding the missing pieces of the "puzzle of light" that Einstein would continue to search for almost another twenty years.

Was he unaware of Boltzmann's method? Not at all. He was very aware of it and had openly criticized its use by both Boltzmann and Planck. Perhaps in 1905 he couldn't bring himself to use Boltzmann's method. Nonetheless, Einstein's 1916–1917 derivation would be as close as anyone would get to a full quantum derivation of Planck's Radiation Law for the next several years, until an unknown physicist from Calcutta, India, revisited the problem in 1924.

Satyendra Nath Bose (1894–1974) was the first and only male of seven children. Bose was an exceptional student of mathematics. In high school,

---

[193] Einstein would have also needed a clear understanding of the photon momentum to easily proceed with calculating the microstates. In 1905, Einstein would have – or at least it's inconceivable to me that he wouldn't have – had a clear perspective on the photon momentum. If he did, he didn't announce it at that time. In fact, the first person to mention the photon momentum was Johannes Stark in 1909, which was the same year (as we discussed earlier) that Einstein showed that the momentum fluctuations of light contain both a photon and wave contribution. Surely, by 1909 Einstein must have been very aware of the momentum of a photon. Yet he said nothing. It wasn't until 1917 that Einstein actually spoke up.

[194] However, as discussed earlier, he had already determined the number of microstates from his previous approach.

he scored 110 out of a total of 100 points on a mathematics exam because in addition to solving all the problems correctly, he solved some using more than one approach. In 1913 and 1915, respectively, he received his bachelor's and master's in mixed mathematics (similar to applied mathematics or mathematical physics today), ranking first in his class on both accounts. In fact, his score on the examination for the master's was so high that it created a new record at the University of Calcutta, which has yet to be surpassed.

At the age of twenty (in 1914), he married Usha Deva. They had nine children. He became a lecturer in physics at the University of Calcutta in 1916. In 1919, Bose, along with former classmate Meghnad Saha (1893–1956), prepared the first English translation of Einstein's Theory of Relativity based on German and French translations of Einstein's original publications. They would continue to collaborate on works in theoretical physics and pure mathematics for several years thereafter. In 1921, Bose joined the physics department as a reader of the University of Dhaka, newly established by the British Imperial Government. It was here that Bose would do the work that would change quantum theory (and his life) forever.

By 1924, Bose had found a way to make Planck's Radiation Law a full quantum theory, free of the assumptions that had been made by Planck and Einstein. It was an amazing accomplishment. Unfortunately, he was unable to get it published on his own, so he decided to write to Einstein:

> Respected Sir, I have ventured to send you the accompanying article for your perusal and opinion. I am anxious to know what you think of it. You will see that I have tried to deduce [Planck's Radiation Law] independent of [classical mechanics] …. I do not know sufficient German to translate the paper. If you think the paper worth publication I shall be grateful if you arrange for its publication in *Zeitschrift für Physik*. Though a complete stranger to you, I do not feel any hesitation in making such a request. Because we are all your pupils though profiting only by your teachings through your writings. I do not know whether you still remember that somebody from Calcutta asked your permission to translate your papers on Relativity in English. You acceded to the request.

The book has since been published. I was the one who translated your paper on Generalised Relativity.

Einstein must have immediately realized what Bose had accomplished, having revisited this problem on many occasions himself – coming very close to solving it in 1917 – only to fall short yet again. Einstein translated the paper for Bose, adding a note of his own to the editor: "In my opinion Bose's derivation of the Planck formula signifies an important advance. The method used also yields the quantum theory of an ideal gas, as I will work out in detail elsewhere." Bose's paper was quickly accepted.

The consequences of Bose's paper are far-reaching, perhaps more so than Bose realized. For nearly two decades, Einstein had essentially been alone in promoting the photon concept (i.e., light quanta, light as a particle). However, in 1924, Bose took it to heart and was able to work a certain kind of magic that had even eluded Einstein.

Bose used the photon concept directly to construct a picture of the microstates in a way that had never been done before.[195] He imagined a given microstate to consist of a collection of little "cells" with a certain number of photons residing in each cell. Therefore, one can imagine that some cells contain 0 photons, others with 1 photon, while others contain 2, 3, and so on. Therefore, a given microstate is described by how many photons are in each of these cells, and as before, the collection of all the microstates describes a macrostate, or thermodynamic state. It's important to note that Bose didn't put a limit on the number of photons that could be in a given cell, nor the total number of photons in the system – any number was possible. In other words, the total number of photons wasn't conserved.[196]

---

[195] Bose's use of the photon momentum directly ($h\nu/c$) played a major role in the development of this new picture of the photon microstates.

[196] In his 1905 treatment, Einstein did assume photons (light quanta) to be conserved. This sneaks into his derivation when he equates the entropy of an ideal gas to the entropy of light, which (as you'll recall) resulted in $E = Nh\nu$, where N is the fixed number of ideal gas atoms. Einstein interpreted N as the number of photons as well, and in doing so implied that photons are conserved.

We've discussed how certain physical quantities in nature are conserved, the most familiar probably being energy. Indeed, energy is always conserved, not being created or destroyed, but merely changed from one form of energy to another. By not making the number of photons of the system conserved, Bose was making a strong physical statement (which, oddly, he never drew attention to in his paper). He was saying that a photon, unlike energy, can be created and destroyed. In fact, we've already seen this in Bohr's model of the atom.

Recall that in Bohr's atom, when an electron "jumps" from a higher to lower quantum state, it emits a photon, whereas absorbing a photon allows it to "jump" up to a higher quantum state. Effectively, emitting a photon is the creation of a photon, and absorbing a photon is the destruction of a photon. Another way to look at it is that a photon is simply a packet of energy, and therefore in the processes of emitting and absorbing, it's merely being transformed from one form of energy to another; the energy is conserved, but the photon isn't.

As did Planck before him, Bose employed Boltzmann's method to determine the total number of microstates for his system of photons and to obtain the resulting entropy.[197] We learned that in Boltzmann's original approach, he imagined each gas atom of his system, "atom 1," "atom 2," etc., as having a specific amount of energy, 0, ε, 2ε, 3ε, etc. For him, a given microstate resulted by labeling each atom and assigning to it one of these specific possibilities for the energy. In this way, Boltzmann's labeling made both the atoms and the energy *distinguishable* (we briefly mentioned this before, but now we'll elaborate). Bose considered his description to pretty much parallel Boltzmann's original approach, but in reality there are striking differences.

Unlike Boltzmann, who focused on the gas atoms and their respective energies, Bose focused on his cells and their respective number of

---

[197] However, Bose sought to maximize the number of microstates to obtain the equilibrium entropy according to Boltzmann's method. On the other hand, Planck merely used Boltzmann's method to obtain the number of microstates, skipping the maximization step, and assumed the result was the equilibrium entropy.

photons. While his cells were labeled ("cell 1," "cell 2," etc.) similar to the way Boltzmann labeled his particles (gas atoms) Bose's particles (photons) weren't. Rather, he only accounted for the number of photons occupying a given cell. So whereas Boltzmann's particles were distinguishable, Bose's particles were *indistinguishable* (another detail not mentioned by Bose).[198] Physically this means that all photons would look the same, if you could actually see a photon up close. They're identical.

Moreover, while Boltzmann's particles were conserved (you can't simply create or destroy gas atoms at your leisure), Bose's particles weren't. Finally, Bose noted that in order to obtain Planck's Radiation Law it was necessary to allow a photon to exist in either of two "polarization states," which in modern quantum theory are known as spin states. This was an amazing find, which Bose noted only in passing.[199]

Identity Crisis

To get a better understanding of distinguishable versus indistinguishable, consider two coins. If we flip each of the two coins and we consider them distinguishable, then the possible microstates are (H_1,H_2), (H_1,T_2), (T_1,H_2), and (T_1,T_2), where H = Head, T = Tails, and the numbers label the coins as "coin 1" and "coin 2." In other words, we have four different microstates. However, if we don't label the coins – that is, we make them indistinguishable – then the possible microstates are (H,H), (H,T), and (T,T), since (H_1,T_2) and (T_1,H_2) are now the same microstate.

In reality, a coin, such as a quarter, isn't really indistinguishable; although two quarters may look "very similar" there will always be some

---

[198] In his theory, Planck treated his resonators as distinguishable, just like Boltzmann treated his gas atoms. However, the energy was treated as indistinguishable, similar to Bose's particles (photons).

[199] Evidently, Bose had discussed with Einstein the idea that photons have a type of spin, but Einstein told him not to emphasize it. The hesitation to elaborate is understandable since the concept of a particle with a "quantum spin" was unknown at that time.

distinguishing characteristic, lending the correct "microstate counting" method to be the former type. On the other hand, microscopic particles such as atoms, photons, electrons, and the like, are indeed indistinguishable and therefore require the latter type of microstate counting method, albeit a much more sophisticated version than the one just described.

It was Bose who introduced such an approach for photons, thereby kicking off what would eventually be known as *quantum statistics*. In 1905, Einstein had asserted a certain equivalence between light and atoms, mainly that light is a kind of particle (photon) just as an atom is a kind of particle.[200] Now some twenty years later, Einstein was ready to extend this equivalence further by applying the method Bose used for light to the atoms of an ideal gas, and in doing so showed that Bose's seemingly benign counting method has dramatic physical consequence for atoms as well.[201] He clearly states this sentiment:

> If Bose's derivation of Planck's radiation formula is taken seriously, then one will not be allowed to ignore [my] theory of the ideal gas; since if it is justified to regard the radiation [light] as a quantum

---

[200] Recall that Einstein compared light to an ideal gas. A consequence of Einstein's assertion is that it implied the microstate counting method for photons was the same as that for an ideal gas. In other words, Einstein used a very different approach from the (correct) method of Bose. Nonetheless, he obtained correct results. This was possible because Einstein only considered light at low density (by using Wien's Radiation Law in his calculation). Thus, when he compared it to an ideal gas, which accurately describes a real gas at low density, it gave correct results. Moreover, the microstates of an ideal gas result from treating the gas atoms as distinguishable. Therefore, Einstein also ended up treating photons as distinguishable. Once again, this worked because Einstein considered each system only at low density.

[201] The papers of Bose and Einstein were really only the beginning of quantum statistics. Today, we understand that with respect to their quantum statistics, microscopic particles fall into either one of two possible groups: they are either *bosons* or *fermions*. Thus, while they are all indistinguishable, they must be further categorized according to their respective group. By applying Bose's method to atoms, Einstein had uncovered the indistinguishable nature of atoms. It would take the work of Paul Dirac (1902–1984), Wolfgang Pauli (1900–1958), and others, which followed shortly thereafter, to clarify the grouping aspect of quantum statistics.

gas, then the analogy between the quantum gas [light] and the molecule gas has to be a complete one.

Einstein wrote three papers dealing with the quantum theory of the mono-atomic ideal gas. In the first of these papers (presented to the Prussian Academy only eight days after Bose's paper was received for publication and published later in 1924), Einstein successfully applied Bose's new method to an ideal gas,[202] obtained the important thermodynamic quantities, and illustrated the difference between his new theory and that of classical mechanics. The pivotal result of this paper is the equivalence it begins to establish between light and atoms. The second paper, which was published in 1925, is the most significant of the three. Here, Einstein directly addresses the *indistinguishability* inherent in Bose's method, which Bose himself never even mentioned. We get a sense that the indistinguishability concept, or "loss of statistical independence," as it was called back then, created a huge upheaval in the physics community. Einstein says:

> An aspect of Bose's theory of radiation and of my analogous theory of the ideal gases which has been criticized by Mr. Ehrenfest and other colleagues is that in these theories the quanta or molecules are not treated as [distinguishable] entities; this matter not being pointed out explicitly in our treatments. This is absolutely correct.

Einstein made no reservations about the inherent indistinguishability. He simply proceeded to detail the differences between the new theory and the traditional classical mechanics approach, and provided a streamlined formula for the counting of indistinguishable particle microstates, which is still used today. Further, he acknowledges that there are real physical consequences stemming from this indistinguishability: "Therefore, the

---

[202] In adopting Bose's approach (in 1924 with his first paper) and applying it to the quantum ideal gas, Einstein had finally accepted Boltzmann's method, whereas earlier he had openly criticized it. This follows from the fact that in calculating the total number of microstates and resulting entropy, Bose made use of Boltzmann's method, albeit a modified version. In following Bose's approach, so did Einstein.

[microstate counting] formula indirectly expresses a certain hypothesis concerning a mutual influence of the molecules of a, at present, totally mysterious kind." Today, we know this mysterious behavior is simply one of the many that surround microscopic particles.

Another striking feature of Einstein's second paper is the prediction of a very unusual phase transition that occurs for the quantum ideal gas. Einstein describes the phenomenon in a letter to Paul Ehrenfest (1880–1933): "From a certain temperature on, the molecules 'condense' without attractive forces, that is, they accumulate at zero velocity."

In other words, as the temperature is lowered, the atoms in the gas begin to "pile up" or condense into the lowest (single-particle) energy state, which is the one with zero kinetic energy; there's a critical temperature whereby a phase transition occurs.[203] This effect becomes most pronounced as the temperature is lowered to absolute zero, at which point all the gas atoms condense into this lowest-energy state. Of course, the idea of a gas condensing into a liquid state is nothing special.

Most of us are familiar with this phenomenon from when our "breath" (the water vapor in our breath, to be specific) condenses when outside on a cold winter day, making it readily visible. The novelty that Einstein was suggesting was that condensation of this sort could occur without attractive interactions of any sort. That is, atoms can be "mysteriously pulled together" to cause condensation, even in the total absence of any sort of attractive interactions (an ideal gas has neither attractive nor repulsive interactions between the particles).

At the time, *Bose-Einstein condensation* (BEC) wasn't taken too seriously and was even met with criticism, particularly from Ehrenfest (which, in part, motivated Einstein's third and least important paper of the series). However, in 1938, Fritz London (1900–1954) suggested that BEC was the

---

[203] It's amazing to note that Einstein found the phase transition through the use of Bose's microstate counting method as applied to an ideal gas. In other words, he identified the phase transition from purely statistical considerations of indistinguishable particles. Nonetheless, today we understand that this type of phase transition is unique to *bosons*.

mechanism involved in the superfluid phase transition of helium-4. Finally, in 1995, Einstein's vision was realized when experimentalists were able to cool a system of rubidium-87 to near absolute zero using a combination of novel cooling techniques.

As if Einstein's clarification of indistinguishability and the identification of a new phase transition weren't enough, he made yet another startling prediction. Once again, Einstein appealed to his fluctuation approach, which he inaugurated in 1904. It had already served him very well, first in his study of light in 1909 and then again in his study of the interaction of light with atoms in 1917. In 1925, it would once again provide amazing insight.

Recall that in 1909, Einstein had successfully implemented this approach to reveal that light behaves both as a particle and a wave. Specifically, he had noted, in his derivation of the energy and momentum fluctuations of light, the existence of both a particle term and a wave term. A similar analysis for the particle fluctuations of the quantum ideal gas also revealed both terms.

In 1909, when Einstein performed this calculation for light, the wave term was expected, while the particle term was surprising. However, in 1925, the situation was reversed: the similar calculation for the quantum ideal gas resulted in the familiar particle term along with an unexpected wave term.

Some twenty years earlier, Einstein had compared light to an ideal gas, leading him to conclude that light could be a particle as well as a wave. It's interesting to note that, at this time, this comparison didn't lead him to ask if an ideal gas atom could behave as a wave as well as being a particle. Apparently, he simply wasn't ready to extend this duality to an ideal gas, and only applied it to light. However, in 1925, by applying Bose's method to an ideal gas, he had shown he was now ready to extend the duality. As it turns out, he wasn't the only one.

Matter Waves

Louis-Victor-Pierre-Raymond de Broglie (1892–1987) was born in Dieppe, Seine-Maritime, France. He was of French nobility, eventually becoming the seventh duc (duke) of the House of de Broglie. Originally interested in humanities, de Broglie graduated with a degree in history. Later his interests turned to mathematics and physics and he received a science degree in 1913. However, his career was cut short with the breakout of World War I. From 1914 to 1918, de Broglie was stationed at the Eiffel Tower as part of the wireless telegraphy section of the army, devoting his spare time to working on technical problems. After the war, he became interested in theoretical physics, particularly quantum theory:

> When in 1920 I resumed my studies ... what attracted me ... to theoretical physics was ... the mystery in which the structure of matter and of radiation was becoming more and more enveloped as the strange concept of the quantum, introduced by Planck in 1900 in his researches into black-body radiation, daily penetrated further into the whole of physics.

De Broglie's older, brother Maurice de Broglie (1875–1960), shared his younger brother's passion for physics. Despite graduating from naval officers' school and spending nine years in the French Navy, Maurice – defying his family's wishes – left it all behind in 1904 to focus on physics, obtaining a PhD in 1908. Beginning his study of X-rays in 1913, Maurice had gained much expertise by 1920, around the time de Broglie began working on his PhD. Although a theoretician at heart, de Broglie maintained quite an interest in the experimental work his brother was doing on X-rays. Indeed, this work along with "long discussions with my brother on the interpretation of his beautiful experiments" made a significant impression upon de Broglie. Foremost, de Broglie became fully convinced that the dual behavior of light that Einstein had described in 1905 applied to X-rays.[204] However, his major conclusion was that this wave-particle duality applied to all quantum particles, especially the electron. In an interview in 1963, de Broglie reflected on his epiphany:

---

[204] De Broglie's application of duality to X-rays was a significant extension of Einstein's 1905 work. At that time, Einstein concerned himself with blackbody radiation in the low-density regime as specifically described by Wien's Radiation Law (not Planck's).

As in my conversations with my brother we always arrived at the conclusion that in the case of X-rays one had both waves and [particles], thus suddenly – ... it was certain in the course of summer 1923 – I got the idea that one had to extend this duality to material particles, especially to electrons.

In 1923, de Broglie (now thirty-one years old) published three papers around this amazing concept, culminating in his PhD thesis *Researches on the Quantum Theory*, which he defended on November 25, 1924. In the first paper, using only Einstein's mass-energy equivalence equation

$$E = mc^2$$

the Planck-Einstein energy equation

$$E = h\nu$$

some very simple mathematics, and his new interpretation of the wave-particle duality, de Broglie derived a formula for the wavelength, λ, associated with a particle,

$$\lambda = \frac{h}{p}$$

where (as before) h is Planck's constant, and p is the momentum of the particle. De Broglie imagined λ to be a wave that accompanied the particle, guiding – or "piloting" – its movements (which he discussed in his second paper), and (in his third paper) he showed that the quantization of an atom, which Bohr had simply assumed, could be arrived at naturally with his pilot wave concept. De Broglie's work had been brought to the attention of Einstein in 1924[205] when Paul Langevin (1872–1946), one of de Broglie's thesis examiners, considering his ideas to be a bit far-fetched (or perhaps downright confusing), sent de Broglie's thesis to Einstein for review. Einstein spoke very favorably of the work:

---

[205] There's some speculation as to if Einstein had already arrived at the concept of wave-particle duality for matter prior to being aware of any of de Broglie's work.

> de Broglie has undertaken a very interesting attempt …. I believe it
> is the first feeble ray of light on this worst of our physics enigmas.
> I, too, have found something which speaks for his construction.

Had it not been for Einstein's enthusiastic response, de Broglie may have not received his PhD for the work. Einstein made good use of de Broglie's concept in 1925, when his analysis of the quantum ideal gas led him to identify (as mentioned earlier) a wave term; it was de Broglie's work (specifically his PhD thesis) that he cited as providing the physical insight.

Earlier we mentioned diffraction is most noticeable whenever a wave passes through an opening smaller than or equal to its wavelength. Indeed, this phenomenon applies to all types of waves. So, if an electron (or any other quantum particle) truly possesses wavelike character, we should be able to observe diffraction by simply passing it through an "opening" that is smaller than or equal to its *de Broglie wavelength*. The question is: What has an opening around the size of the de Broglie wavelength of an electron?

It turns out that the tiny spaces (the atomic spacing) between atoms that form a crystal solid are ideal. In other words, a crystal solid with the right atomic spacing should act as a diffraction grating for electrons. In fact, there was already precedent for this based on the work of Max von Laue (1879–1960), who in 1912 showed that a copper sulfate crystal diffracts X-rays, thereby proving their wave nature (at the time the nature of X-rays was still heavily debated).

Confirmation of de Broglie's theory and the wave-particle duality came in 1927, when Clinton Davisson (1881–1958) and Lester Germer (1896–1971) successfully showed that a beam of electrons directed at a nickel crystal results in more than their simply being deflected off the nickel; it causes the electrons to diffract. This experiment was a culmination of a decade's worth of effort, which originated with Davisson.

The experiments were conducted in a vacuum chamber to avoid adverse effects from air. However, an accident occurred that allowed air to enter the chamber, resulting in the nickel reacting with the oxygen (in the air) to produce nickel oxide. To reverse this *oxidation* of the nickel, they heated it in the presence of hydrogen. However, after resuming their experiments with their "restored" nickel, they obtained some interesting new results. The amount of deflection of the electrons from the nickel was more in some directions than others. In other words, there was a directional dependency to the electron intensity that hadn't been seen in their previous experiments.

They speculated that the newly restored nickel had actually changed from its original solid form into a new crystal form. Further, they reasoned the crystal form had the needed atomic spacing to diffract – rather than simply deflect – the electrons, thereby revealing their "de Broglie" wave nature. Finally, they found that *Bragg's Law* correctly described the directional dependency of the intensity they had seen in their experiments.

I won't go over the details of Bragg's Law, but the main point is that it correctly describes the directional dependency of the intensity under the assumption that what's being deflected is a wave – not a particle. In other words, it correctly describes diffraction of waves by a crystal (von Laue's X-ray diffraction experiments are correctly described by Bragg's Law). Evidently, Davisson and Germer had no intent of testing de Broglie's theory, but were simply interested in investigating the surface properties of nickel.

Up until 1926, they were unaware of de Broglie's theory, and it wasn't until Davisson was attending a scientific meeting that it was brought to his attention. He and Germer were already finding that their experimental data was similar to the results obtained for X-ray diffraction of crystals. With the insight provided by de Broglie's theory, Davisson now realized this was due to the wave nature of the electron.

The experiments of Davisson and Germer were a big step forward for the wave-particle duality of matter, in particular the electron. Particles and waves had always been considered separate entities in classical physics, but for quantum entities (photons, electrons, etc.) this distinction disappears and they become intertwined.

In 1929, de Broglie received the Nobel Prize for his theory of wave-particle duality. He was only thirty-one years old (in 1923) when he wrote his groundbreaking series of papers. The concept of the de Broglie wavelength would be his one great contribution to physics. However, it would be instrumental in inspiring Erwin Schrödinger – a thirty-eight-year-old physicist

– to develop the equation (while on Christmas break with his mistress) that is the very heart and soul of what's now known as quantum mechanics.

Schrödinger's Path to Quantum

Erwin Schrödinger (1887–1961) was born in Vienna, Austria, the only child of Rudolf and Georgine Schrödinger. His father inherited the family business, a successful linoleum and oilcloth production factory, and his mother descended from a minor nobleman. Indeed, the Schrödingers were well off and enjoyed a lifestyle typical of the upper-middle class of the time. Schrödinger grew up surrounded by women (two aunts, a female first cousin, nurses, and maids) catering to his every need, which most likely influenced his later relationships with women. And for some time, not even school disrupted this comfortable setting, as Schrödinger was privately tutored until he was ten.

However, it would be his father who undoubtedly provided Schrödinger with his greatest intellectual stimulation initially. Although his father ran the family business, he was a scientist at heart with his true passion being botany (a subject he published on in professional journals). He was also very knowledgeable in Italian painting and chemistry. To pursue his interest, he maintained a huge library, which was readily available to his son, whose education was of the utmost priority. Later in Schrödinger's life, it would be his father who discouraged him from taking up the family business and to instead pursue an academic career (something the senior regretted not doing himself).

After a long holiday in the spring of 1898, Schrödinger got his first taste of formal education at St. Nicklaus School, where his worried parents sent him for just a few weeks in preparation for his entrance exam to the Gymnasium (similar to an English grammar school). Schrödinger passed with flying colors and entered the Gymnasium in the fall of 1898, which was conveniently located ten minutes from his home. The major subjects at the Gymnasium were Latin and Greek languages and literature, with the minor subjects being mathematics and physics. Schrödinger did very well and was always at the top of his class in every subject.

In 1906, shortly after Boltzmann died, Schrödinger entered the University of Vienna. Boltzmann's death left the physics department in disarray with the theoretical physics lectures being suspended for eighteen months until his replacement, Friedrich (Fritz) Hasenöhrl (1874–1915) arrived. Nonetheless, it would be worth the wait, as Hasenöhrl was up and going in no time, giving a masterful debut lecture on Boltzmann's statistical mechanics (he himself had studied under both Boltzmann and Josef Stefan (1835–1893) at the University of Vienna). Hasenöhrl's course on theoretical physics was five days a week spread over eight semesters – Schrödinger couldn't have been happier: "No other person has had a stronger influence on me than Fritz Hasenöhrl, except perhaps my father."

Theoretical physics was to be his calling, and in 1910 Schrödinger received his Doctor of Philosophy (not quite equivalent to a PhD in America, more like a master's) for his dissertation *On the Conduction of Electricity on the Surface of Insulators in Moist Air*, motivated by the significance of the electrical insulation used in instruments to measure ionization and radioactivity.

Soon afterwards, Schrödinger was called up for military service. He had signed up in 1908 to be part of the fortress artillery. Although military service was a three-year requirement for all able-bodied men, since there was a need for more officers than the academies and cadet schools could provide, the system allowed for men of adequate education and social standing – such as Schrödinger – to be "one-year volunteers" who trained as officer candidates, covering their own living expenses. After completing his military training, he passed the examinations for the rank of Fähnrich in the reserves (a rank just below lieutenant for cadet officers) and returned to full-time civilian life on New Year's Day 1911.

Returning to the university, Schrödinger became an assistant to the experimental physicist Franz Exner (1849–1926), a more suitable position (and one for which he would have surely been chosen for) with Hasenöhrl having passed him by due to his military service. Schrödinger was now entrenched in the habilitation process, which would move him to the first rung on the academic ladder as *Privatdozent*. The process involved several hurdles, with the first being the completion and publication of original scientific research.[206] Next was a formal presentation to the committee known as *Kolloquium.*

As required, Schrödinger chose three possible topics for the presentation: anomalous dispersion in the electric spectrum; the significance of the quantum of action in the theory of heat radiation; and the magneton, which the committee chose for discussion. Finally, there was an oral exam and discussion on the topic of *Kolloquium* before the committee. Having fulfilled all the requirements, on January 9, 1914, Schrödinger became a *Privatdozent.* On July 31, 1914, his father arrived at Schrödinger's office to personally deliver his mobilization efforts.

Schrödinger was now officially part of the war effort and would be until the end in 1918. Schrödinger would see most of his war activity as acting commander of battery during the third battle of the Isonzo[207] (October 18 – November 3, 1915). His distinguished service there resulted in a military citation, which is on his record in the War Archive as follows:

> while acting as a replacement for the battery commander, [Schrödinger] commanded the battery with great success. During the preparations as well as in many engagements, he was in command as first officer at the gun emplacement. By his fearlessness and calmness in the face of recurrent heavy enemy artillery fire, he gave to the men a shining example of courage and gallantry. It was owing to his personal presence that the gun emplacement always fulfilled its assignment exactly and with success in the face of heavy enemy fire ....

On May 1, 1916, Schrödinger was promoted to *Oberleutnant.* In 1917, respite came when Schrödinger was sent back to Vienna to teach a laboratory

---

[206] Out of this effort came the papers *On the Kinetic Theory of Magnetism,* and *Studies on the Kinetics of Dielectrics, the Melting Point, Pyro- and Piezo-Electricity.*

[207] Today, the Isonzo is located in present-day Slovenia. It's a sixty-mile-long river valley flanked by mountains on each side, flowing north-south from the Julian Alps to the Adriatic Sea. During World War I, it was located just inside Austria-Hungary along its border with Italy. There were twelve battles of the Isonzo between 1915 and 1917.

course in physics at the university, along with a meteorology course at a school for antiaircraft gunnery officers. However, things weren't good for long.

In 1918, Schrödinger's mother was recovering from a breast cancer surgery, which was performed almost a year earlier, and Schrödinger himself struggled with tuberculosis. And the family business, unable to obtain the needed raw materials, finally collapsed. Schrödinger's father was also ill, showing signs of hypertension and atherosclerosis, and on December 24, 1919 – while resting in his chair – he died.

Fortunately, by 1920 Schrödinger's job prospects had improved – just in time, as the family savings were now completely gone. He was first offered a promotion to associate professor at the University of Vienna. This prospect was rather dismal, as the salary wouldn't provide enough to actually support him and his soon-to-be wife, Annemarie (Anny) Bertel. Luckily, around this time he was also offered an assistantship (which also included some lecturing duties on modern theoretical physics) at the university in Jena. The salary was an improvement over the previous offer, and so he accepted. Before they left, Schrödinger and Annemarie got married (twice, actually) and arrived in Jena in April that year.

He made an excellent impression, and after only a few weeks he was recommended for associate professor by the faculty. Although the salary significantly exceeded the current one, the post-war inflation was on the rise, and the position wasn't a permanent one. So, when a regular associate professorship in Stuttgart was offered, he gladly accepted, resigning from his position at Jena on October 1. Nonetheless, his stay in Stuttgart wouldn't last long, either.

In the early summer of 1921, he and Anny moved to Breslau (now Wrocław), where Schrödinger accepted yet another position, this time as full professor. Prior to moving, Schrödinger had confided in a friend about the post-war tension that was prevalent in the region: "I often have the feeling: let me get away from this powder keg." Less than six months later, the Schrödingers moved again.

The chair of theoretical physics at the University of Zürich had remained vacant all through the war, although the department had exceptional lecturers at its disposal, in particular Paul Epstein (1883–1966). The Swiss government was, well, cheap, and not making the appointment anytime soon was saving them money. By November 1919, the faculty had had enough and requested that the position be finally filled. This process dragged on, but on March 3, 1921– more than a year later – the dean wrote to the Education Commission that Laue should be the first choice.

Laue had indicated his interest in coming back – he had held this position before – but made it clear he would need an exceptionally good salary to do so, given his financial crisis (his family had lost all their money in the post-war inflation). Apparently, the directorate couldn't come up with enough money to meet Laue's salary requirements.

About a year earlier, the committee had feared it wouldn't be able to attract top talent and had considered their (three) lecturers. Epstein had been strongly considered, but there were concerns that his Polish accent would make teaching the large-sized elementary lectures an issue. Apparently this was not an issue for the University of Leiden, which offered him a position with Hendrik Lorentz (1853–1928), which he gladly accepted.

Now with both Laue and Epstein unobtainable, the committee wasted little time and, on July 20, 1921, officially approved Schrödinger's appointment as full professor of theoretical physics. At roughly $2,500 per year, Schrödinger's salary was top-notch for the times. Moreover, as successor to Einstein, Peter Debye (1884–1966), and Laue before him, he was following in some very eminent footsteps. Finally, after moving from one position to another and with constant financial worries, the thirty-four-year-old Schrödinger must have been both relieved and inspired. Schrödinger would remain at the University of Zürich for almost six years. Here he would do his greatest work, formulating the very equation that was pivotal in transforming quantum theory into quantum mechanics.

Indeed, he might have been inspired, but Schrödinger was off to a rough start at his new position. No sooner did he begin the fall semester in 1921 than everything came to a grinding halt: in November Schrödinger – now both physically and mentally exhausted from a series of events since 1918 (the loss of both parents and a grandfather and economic difficulties) – suffered from tuberculosis compounded with a severe case of bronchitis. With no antibiotics available at the time (the first true antibiotic, penicillin, was discovered in 1928), the only treatment for tuberculosis was taking rest at a high altitude. Schrödinger took to the alpine resort Arosa. He remained there until November 1922, returning a bit late for the following academic year. It wouldn't be his last visit, as he would regularly return to resolve health issues or simply for vacation. Remarkably, despite his weakened state, Schrödinger was able to publish two papers while recovering in Arosa. The first was nothing special, but the second hinted at what would culminate four years later in his most significant work ever.[208]

---

[208] In 1916, Einstein finalized general relativity. He showed that in the presence of matter, spacetime becomes curved, resulting in the "force" we call gravity. And so, gone were the days of thinking of gravity as a force acting (instantaneously) at some distance to draw one object towards another. In 1918, Hermann Weyl (1885–1955) attempted to unify the only two forces known to physicists at the time, gravity and electromagnetism, under the framework of Einstein's general relativity by introducing a gauge transformation. In Einstein's (Riemannian) spacetime the magnitude of a vector remains constant, or invariant, as it moves from point to point along a spacetime trajectory. However, with Weyl's gauge transformation, all this changes, and the magnitude of a vector moving along a trajectory in this new (non-Riemannian) spacetime will now vary. The good news was that, from a purely mathematical perspective, Weyl had successfully unified gravity and electromagnetism. He sent his results to Einstein, who was initially excited but ultimately was unable to accept the actual physical consequences of Weyl's approach. Embodied in Weyl's theory was the fact that the length of a measuring stick will vary from point to point along a spacetime trajectory. In other words, a measurement taken at one point would have a different value when taken again at another point, simply because of the location change in spacetime. Moreover, the rate of change of time will also vary from point to point along a spacetime trajectory. In other words, length and time measurements are not absolute anymore (as they were in Einstein's theory). Instead, they are now relative and depend on the location they're measured at – they are locally dependent.

In particular, Einstein pointed out that according to Weyl's theory, the atomic spectra of a given element would depend on where and when the measurement was taken. However, we know this is not the case physically. In 1922, in his paper *On a Remarkable Property of the Quantized Orbits of a Single Electron*, Schrödinger revisited Weyl's approach. He considered an electron in orbit around a hydrogen atom according to Bohr's atomic model. He further imagined an associated vector,

Schrödinger returned to a heavy schedule of eleven hours a week of lecturing; as a senior professor he was expected to shoulder a heavier load than junior faculty – a scenario quite the reverse at most of today's universities. His heavy lecture schedule for the academic year of 1922–1923, along with his still-weakened health, made 1923 a low point for publications – he published none. This is rather curious, considering he published during his previous hiatus and while a soldier in World War I.

In 1924, Schrödinger attended the 88th Meeting of German Scientists and Physicians, held on September 21–27. There were many in attendance, with important talks by prominent scientists such as Arnold Sommerfeld (1868–1951), Max Born (1882–1970), Max Planck, Paul Ewald (1888–1985), Max von Laue, and several others. Although Einstein didn't give a presentation, his light quanta concept – now almost twenty years after its inception – was a hot topic of debate among the smaller group of attending experts. With Compton's work of 1923, many physicists had become believers of light quanta, but there were still a few stragglers. The most notable of these were Niels Bohr, Hendrik Kramers (1894–1952), and John Slater (1900–1976).

In 1924, the trio wrote a paper that outright rejected light quanta. In fact, so much were they against it that they were willing to toss out the

---

whose magnitude varied in accordance with Weyl's theory, as the electron moved (through spacetime) from point to point in its orbit. Exactly what physical property the vector represented was unclear to Schrödinger at the time. However, he noted that mathematically (with the correct choice for the undetermined constant, which he related to Planck's constant), it was possible to preserve the magnitude of the vector over the electron's orbit, thus avoiding any physically undesirable effects. In other words, it appeared that in this circumstance Weyl's theory was physically realizable. Although he was unable to determine the physical significance of the presumed vector, Schrödinger remarked, "It is difficult to believe that this result is merely an accidental mathematical consequence of the quantum conditions, and has no deeper physical meaning." Later, Schrödinger would realize that the physical significance of his undetermined vector was none other than de Broglie's wavelength. It's interesting to speculate that had Schrödinger not been so worn out, he might have been able to turn this inkling into his greatest work at that time rather than four years later.

conservation of energy (the first law) and the conservation of momentum in their attempt to eliminate light quanta:

> As regards the occurrence of [electrons jumping between quantum states], which is the essential feature of the quantum theory, we abandon ... a direct application of the principles of conservation of energy and momentum ....

In other words, rather than strict conservation for each individual jump, they instead suggested that the conservation of energy and momentum should be treated as holding only on average over many electron jumps; conservation only occurs in a statistical rather than absolute sense. The paper, with all its implications, became known as the BKS proposal (after their last names). It's interesting to note that Schrödinger himself had expounded on the same notion (probably influenced by his time working with Exner) in his 1922 inaugural address at the University of Zürich:

> It is quite possible that the laws of nature without exception have a statistical character. To postulate an absolute law of nature behind the statistical one, as is generally done today as a matter of course, *goes beyond the bounds of experience. ... The burden of proof lies upon the advocates of absolute causality, not upon those who doubt it.* For to doubt it is today by far the more *natural* viewpoint.

It's no surprise then that for the most part, Schrödinger was in favor of the BKS proposal. In contrast to the BKS proposal stood Bose's recent paper, in which he successfully used light quanta to provide the first fully quantum derivation of Planck's Radiation Law. In turn, Einstein had just successfully applied Bose's quantum statistics in his derivation of the quantum ideal gas. Then there was Planck, who had his own theory of the quantum ideal gas, which he originally constructed in 1916, and later lectured on in a series given at the University of Munich in the winter of 1924–1925. He gave a taste of his approach to quantum statistics as applied to hydrogen gas in his presentation at the meeting. Indeed, the new field of quantum statistics was a likely conversation point at the meeting, and one can imagine that its importance would have been impressed upon Schrödinger.

Unfortunately, whatever thoughts might have been sparked in Schrödinger's mind would have to wait since his 1924–1925 winter semester teaching duties left little time for his own research. Things changed towards the end of the semester, and now Schrödinger was intensely studying the literature on quantum statistics. In addition to color theory, Schrödinger's primary area of research during his initial years in Zürich was Boltzmann's statistical approach to gas theory, which, as we've seen, was the precursor to quantum statistics.

Given this expertise and his statistical view of nature at the atomic level (as evident from his 1922 inaugural lecture), quantum statistics must have had great appeal for Schrödinger. On February 5, 1925, Schrödinger wrote to Einstein in Berlin, "Just now I have read your interesting [first paper on the quantum ideal gas], and I have encountered a serious [issue]." He goes on to say, "I therefore very much doubt whether one can ascribe any real significance to the deviations [you're noting in your calculations] from the [classical] behavior of an ideal gas."

It seems that upon reading Einstein's paper, Schrödinger had missed one of the main points: quantum particles are indistinguishable, which means that (in general) they don't follow Boltzmann's statistics, but rather the new quantum statistics of Bose and Einstein.

Einstein politely replied: "I have not made a mistake in my paper. In Bose statistics, which I use, the [light] quanta or molecules are not considered as being *mutually independent* objects." Einstein even went on to schematically illustrate how the microstate counting method of Bose differs from that of Boltzmann.

By the summer semester of 1925, Schrödinger was now lecturing on quantum statistics. This, along with his previous exchange with Einstein, probably motivated him to further review the current literature around the topic to tighten his understanding. Finally, On November 3, 1925, he responded to Einstein:

> Only through your letter [from February 28] did the uniqueness and originality of your statistical method of calculation become clear to me; I had not grasped it before at all in spite of the fact that Bose's paper had already come out. ... Only your theory [on the quantum ideal gas] is really something fundamentally new, and this I did not grasp at all, at first.

Ironically, although Schrödinger now appreciated the significance of Einstein's use of Bose's counting method, he seemed to consider Bose's original application to light quanta somewhat inconsequential. Also, in this letter Schrödinger expressed his enthusiasm for an idea Einstein had mentioned in another letter back in September. Einstein had reflected on a fundamental approach in Planck's theory of the quantum ideal gas, whereby one treats the energy levels of the entire collection of atoms, rather than focusing on the single-particle energy levels, which (as we saw) formed the basis for both Bose's and Einstein's theories.

Just two days later, Schrödinger had followed through with Einstein's suggestions, writing to Einstein, "The basic idea is yours, I have only carried out the calculation .... I need not emphasize the fact that it would be a great honor for me to be allowed to publish a joint paper with you." Einstein graciously declined authorship, not wanting to be an "exploiter" of the work Schrödinger had done. Nonetheless, Einstein's contribution was duly noted in the introduction of Schrödinger's paper (most likely completed in early December).

In the end, the paper provided an opportunity for Schrödinger to refine his expertise in quantum statistics. His resulting formulas were of no practical use, and rather than reconciling the quantum ideal gas theories of Einstein and Planck, Schrödinger had now introduced a third into the mix, which more closely resembled Planck's than Einstein's.

Nonetheless, his efforts had made him well versed on the current literature and provided him with expertise in quantum statistics, neither of which was the case prior to his initial contact with Einstein back in February. He was beginning to connect the pieces.

The Wave Equation

On November 3, 1925, Schrödinger wrote to Einstein:

A few days ago I read with the greatest interest the ingenious thesis of Louis de Broglie, which I finally got hold of; with it [the discussion on the wave nature of the quantum ideal gas in your second paper] has also become clear to me for the first time.

As mentioned earlier, in his second paper on the quantum ideal gas, Einstein performed an analysis of the particle fluctuations and found that the resulting expression gave both the expected particle term along with a very unexpected wave term. Einstein declared, "this concerns more than a mere analogy" and noted the wave-particle duality outlined by de Broglie in his thesis: "How a [wave] can be assigned to a material particle or a system of material particles has been pointed out by Mr E. de Broglie in a very noteworthy treatment." (Einstein mistakenly refers to de Broglie's first initial as "E" rather than "L".) Schrödinger wasn't kidding when he told Einstein things had become clear.

On December 4, 1925, Schrödinger notified Einstein that he had submitted the paper on quantum statistics. He then proceeded to outline the goals for his follow-up paper. This would be his last paper before his most famous work, which changed all of physics as it was known. In short, Schrödinger wanted to develop a theory of the quantum ideal gas that was completely free of Bose statistics with its implication that atoms reside in little "cells" as employed in Einstein's work. He simply didn't accept the new statistics as something fundamental. Rather, he argued that Einstein's approach was actually concealing the deeper underlying physical meaning.

In order to accomplish his goal, he (once again, albeit with a different approach) treated the energy levels of the entire collection of atoms, rather than focusing on the single-particle energy levels as Einstein had done. Schrödinger noted that Bose statistics led to both Einstein's quantum ideal gas theory and to Bose's light quanta theory of Planck's Radiation Law. Further, he pointed out that Planck's Radiation Law can also be obtained by treating light according to its wavelike character.

Clearly, the wave-particle duality at the quantum level was compelling. Thus, in order to complete the wave-particle duality picture, Schrödinger proposes a new theory where a system of gas atoms is treated as waves using, rather than Bose statistics, "natural statistics" (a statistical approach with experimental validation and, in Schrödinger's opinion, more logically grounded):

> Hence one must construct the picture of the gas simply according to *that* picture of [blackbody radiation], which does *not yet* correspond to the extreme light-quantum concept; then the natural statistics ... will yield Einstein's gas theory.

He added, "This means nothing more than taking seriously the [proposed theories of de Broglie and Einstein, where atoms are treated according to their wave nature]."

At this point, Schrödinger was fully committed to developing a complete theory that treated the atoms of the ideal gas in terms of their wavelike behavior as originally described by de Broglie and further investigated by Einstein. It's interesting to note that neither de Broglie nor Einstein had successfully accomplished such a theory. It's reasonable to assert that de Broglie simply lacked the mathematical prowess given his formal background. A bit more surprising is that Einstein didn't. Indeed, he had been the father of wave-particle duality with his 1905 paper on light quanta. Although it had been Bose who successfully used Einstein's light quanta concept to obtain the correct quantum statistics and finally put Planck's Radiation Law on solid quantum ground, Einstein was the one who immediately implemented the "duality principle" to successfully apply Bose's statistical method to the ideal gas.

Further (as already mentioned), Einstein was well aware of how to use de Broglie's method to represent a system of gas particles as a wave and had considered such an approach (in his second paper on the quantum ideal gas) to solve a well-known problem. In the end, it was Schrödinger who showed a collection of ideal gas atoms could be described with a wave approach to obtain correct results, just as Einstein had done using Bose's "unnatural" statistical method. And while it was an admirable piece of work, Schrödinger had, for the most part, simply tied up the loose ends left behind by Einstein and de Broglie, while still leaving some key components to his wave theory missing.

Early in November 1925, Debye suggested Schrödinger give a talk on de Broglie's work at the joint University of Zürich/ETH colloquium. Apparently, according to Felix Bloch (1905–1983), the invite went something like this:

> Schrödinger, you are not working right now on very important problems anyway. Why don't you tell us some time about that thesis of de Broglie, which seems to have attracted some attention?

The colloquium was an informal lecture series (occurring every two weeks for two hours each time) with an audience of perhaps a couple of dozen at the most. Although not known for sure, Schrödinger most likely presented between the second half of November and the first half of December, and probably around the same time he was working on his initial "wave theory" (second quantum statistics) paper. Schrödinger gave a beautiful overview of de Broglie's work. Nonetheless, Debye wasn't satisfied and felt there was still something missing from de Broglie's approach. Once again, according to Bloch, "Debye casually remarked that he thought this way of talking was rather childish. As a student of Sommerfeld he had learned that, to deal properly with waves, one had to have a wave equation."

Debye had a good point. In classical physics, all waves have an associated wave equation. Simply put, a wave equation describes the physical motion of the wave through space and time; it's to waves what Newton's equation of motion (Newton's Second Law) is to particles. It's not clear how much of a role Debye's comment really had in getting Schrödinger to move forward on developing a wave equation. Most likely, it got him thinking more seriously about it, even though it's clear he was already heading in this direction. In fact, one could easily argue that a wave equation was already implicit in his original wave theory, albeit far from fully developed.

Undoubtedly, a wave description of the quantum world was much more appealing to Schrödinger than the "electron hopping" picture described by

Bohr's atom. Following closely from the work of de Broglie, Schrödinger attempted to develop a wave equation of an electron moving in the hydrogen atom, the simplest of all atoms. Naturally, he included the effects of special relativity in his equation. However, he hit an obstacle and found that his new relativistic equation failed to describe well-known observable effects. Schrödinger never published these results, probably discouraged by the theory's inaccuracies. Nonetheless, he wouldn't be stumped for long.

Before Christmas break, Schrödinger wrote "an old girlfriend in Vienna" asking her to join him in his favorite getaway place, Arosa. He had spent the previous two Christmases there with his wife, Anny, but this time was looking for a change of pace, apparently. The identity of this woman remains a mystery, but whoever she was, it seems she served as an amazing source of inspiration for him.

Schrödinger, now thirty-eight, began what would become a twelve-month period of sustained creativity throughout 1926, which resulted in six major papers on a new theory of quantum known as *wave mechanics*. Shortly after returning from Arosa, Schrödinger gave another colloquium, in which he opened with these words (according to Bloch): "My colleague Debye suggested that one should have a wave equation [for the electron in the hydrogen atom]; well, I have found one!"

Yes, Schrödinger had found the wave equation. Unfortunately, he wasn't able to solve it yet, and he appealed to some colleagues for help. As to how much help he received, or who even helped him, it's unclear. Schrödinger was a very capable mathematical physicist in his own right, having received some very excellent training. Nonetheless, it appears Schrödinger's close friend and mathematician Hermann Weyl (1885–1955) – who happened to also be the lover of Schrödinger's wife – provided some assistance, which Schrödinger clearly acknowledged.

Less than three weeks had passed since his return from Arosa when the journal received Schrödinger's first paper on January 27, 1926, titled *Quantization as an Eigenvalue Problem*. Here he outlines the process[209]

---

[209] The reality is that he already knew the wave equation before providing this "derivation," having most likely obtained it from a few very simple mathematical

that led him to the time-independent wave equation, using the (nonrelativistic) hydrogen atom as an example. Now, with the wave equation in place, the goal becomes to find its solution, which is given by the *wavefunction*.

He successfully finds the wavefunction along with the electronic energy levels for the hydrogen atom, whereby he correctly obtains the Balmer Formula. Recall that in his atomic theory, Bohr didn't actually provide a derivation for the Balmer Formula. Rather, he assumed it's the correct formula for the hydrogen spectrum and then used it to obtain the electronic energy levels. Moreover, Balmer himself never provided a formal derivation; it was something he simply "came up with." Towards the end of the paper, Schrödinger acknowledges his debt to de Broglie: "Above all, I wish to mention that I was led to these deliberations in the first place by the suggestive papers of M. Louis de Broglie ...."

He then goes on to mention his second quantum statistics paper, in which he provided a "wave theory" for Einstein's quantum ideal gas, and how his current work can be considered a generalization of that work. This is a telling statement, showing that his prior work in quantum statistics was the rightful predecessor of wave mechanics. The next paper was received just four weeks after the first on February 23, and the other four followed very quickly, with the last one sent on June 21.

The reactions to Schrödinger's wave mechanics came shortly thereafter. Planck commented that he had read through the first paper "like an eager child hearing the solution to a riddle that had plagued him for a long time." Regarding the second paper, Planck was once again enthusiastic: "You can imagine with what interest and enthusiasm I plunged into the study of this epoch-making work ...." Planck was indeed impressed and was looking to bring Schrödinger in as his successor when he retired in 1927.

---

manipulations. You can't directly derive Schrödinger's wave equation from classical mechanics. This is because, in addition to the mathematical formulas, specific postulates are required. In particular, Schrödinger required the wavefunction of his wave equation to be real, single-valued, finite, and continuously differentiable up to second order. By his fourth paper, he would conclude that, sometimes, the wavefunction is a complex (i.e., not real) quantity.

Einstein weighed in, saying, "the idea of your work springs from true genius!" Ten days later Einstein added, "I am convinced that you have made a decisive advance with your formulation of the quantum condition …."

Ehrenfest wrote:

I am simply fascinated by [your theory] and the wonderful new viewpoints that it brings. Every day for the past two weeks our little group has been standing for hours at a time in front of the blackboard in order to train itself in all the splendid ramifications.

For the most part, everyone was excited by Schrödinger's wave mechanics. However, it wasn't the only version of quantum mechanics around.

Two Quantum Mechanics: Schrödinger and Heisenberg

Over a decade had passed since Bohr introduced his atomic quantum theory of electrons jumping to lower (electronic) energy levels, or orbits, that emit a photon of a specific frequency in the process. Bohr's atomic model provides a very nice, physically intuitive picture. But the reality is that we can't actually measure the motion of an electron in an orbit, or even truly know if this motion is real. On the other hand, the frequency spectrum (spectral lines) resulting from the electron transitions to lower energy levels is something we can see, and after all this time it still provided the only clues as to the inner workings of the atom.

It was Werner Heisenberg's (1901–1976) argument that a proper quantum theory should only include physical quantities that are actually observable. At just twenty-three years old, he set out to construct such a theory. By late May 1925, after some initial progress, Heisenberg was struck with a severe case of hay fever, which forced him to take a break from his regular routine:

I fell so ill with hay fever that I had to ask [my professor] Born for fourteen days' leave of absence. I made straight for Helgoland, where I hoped to recover quickly in the bracing sea air, far from blossoms and meadows.

Apparently the isolation of this small North Sea Island not only cleared his hay fever, but also his mind. He says:

> Apart from daily walks and long swims, there was nothing to distract me from my problem, and so I made swifter progress than I would have done [back home at the university of] Göttingen.

We won't go over all the details of Heisenberg's theory here, but will rather focus only on the important aspects. Heisenberg's theory contains two important features. The first one is the complete set of frequencies emitted by an atom due to electrons jumping to lower energy levels. The second component he included is the probabilities associated with these electronic transitions.

A major concern Rutherford had with Bohr's atomic model was its lack of deterministic behavior:

> There appears to me one grave difficulty in your hypothesis, which I have no doubt you fully realize, namely, how does an electron [that is about to make a transition] decide what [energy level] it is going to [transition to as it] passes from one [energy level] to the other? It seems to me that you would have to assume that the electron knows beforehand where it is going to stop.

By including the set of transition probabilities, Heisenberg had effectively resolved this issue by affirming that the quantum realm doesn't have the familiar determinism of classical physics that Rutherford desired. Rather, the various transitions simply occur according to their probability. There was already precedent for this from Einstein's work of 1916–1917, where he had used the concept of transition probabilities for atoms interacting with light.[210]

---

[210] Einstein actually considered three processes: spontaneous emission, stimulated emission, and absorption. Heisenberg focused only on spontaneous emission, whereby a transition is made without interacting with light. It was spontaneous emission that Einstein drew attention to at the end of his paper, noting that from

By the time Heisenberg returned from Helgoland, he had his version of quantum mechanics in hand and immediately began writing up his results, although not without hesitation, as he confided in a letter to his father, "My own works are at the moment not going especially well. I don't produce very much and don't know whether another [paper] will jump out of this at all in this semester."

In early July 1925, hesitantly Heisenberg went to Born with his paper asking him to look it over and decide if it was publication-worthy. Upon reading the paper, Born was struck by the mathematical entities arising from Heisenberg's theory and was sure that he had seen them somewhere before. Finally, he recalled in a letter to Einstein, "Heisenberg's [mathematical formulation] did not give me rest, and after days of concentrated thinking and testing I recalled an algebraic theory I had learned from my teacher ...."

The theory Born was referring to was that of matrices, specifically, the rules for multiplying two matrices together. Heisenberg had been totally unaware that he had ended up deriving this fundamental matrix relation in the course of his new theory. Since matrices weren't common knowledge among physicists at that time, Born would have been one of very few physicists who could have actually recognized it. At the end of the month, Born forwarded Heisenberg's paper on to a journal for publication.

Heisenberg's new quantum mechanics was presented in the paper titled *Quantum Mechanical Reinterpretation of Kinematic and Mechanical Relations*. Immediately, Born and his young assistant, Pascual Jordan (1902–1980), extended and clarified Heisenberg's work in a paper sent off for publication in September 1925. All three of them came together for the third ("three-man") paper, submitted in November 1925. With this final effort, they proclaimed they had found the long-awaited-for theory of quantum mechanics.[211] And all this was done before Schrödinger's first paper was

---

his theory the direction of the momentum recoil for this process was – much to his chagrin – solely determined by chance.

[211] It's these last two papers that form the foundation of the matrix mechanics approach to quantum mechanics as taught today.

received by the journal (on January 27, 1926). Of course, Schrödinger had known about the first two papers, but not the third, when he formulated wave mechanics. Regardless, they had no impact on his work.

This is because from a mathematical standpoint, the two approaches are very different. Central to Schrödinger's approach is the wave equation, a partial differential equation, whose solution is given by the all-important wavefunction. Heisenberg's formulation has nothing to do with a wave equation or partial differential equations; rather, it involves matrices, which is why it's often referred to as *matrix mechanics*. Further, his formulation by design contains discontinuous energy levels and their associated transition probabilities, whereas in Schrödinger's wave mechanics, the energy levels are not built in – they simply come out naturally upon solving the wave equation and applying the correct boundary conditions to the continuous wavefunction solution.

As for the probabilistic nature present in Heisenberg's theory, well, it's explicitly absent from Schrödinger's formulation. This may seem rather odd to you since we noted on several occasions that probability seems to be inherent in quantum mechanics. We'll see in a moment that Schrödinger's theory does have this inherent probability as well; it's just a bit hidden at first. So on the surface these two theories seem very different, not only mathematically, but also in their physical description, a fact their respective proponents enthusiastically emphasized.

Schrödinger stressed the continuous (essentially classical mechanics) nature of his theory, imagining the electronic transitions in atoms associated with the emission and absorption of light to occur – not through the electrons jumping as Bohr had originally described – but through smooth transitions whereby a vibrating electron (something similar to one of Planck's resonators) simply "transitions" from one vibrational state to another:

> It is hardly necessary to emphasize how much more congenial it would be to imagine that at a quantum transition the energy changes over from one form of vibration to another, than to think of a jumping electron.

Schrödinger made no secret of his dislike for Heisenberg's theory with its matrices, jumping electrons, and lack of an intuitive physical picture:

> I naturally knew about [Heisenberg's] theory, but was discouraged, if not repelled, by what appeared to me as very difficult methods of [matrix mechanics], and by the want of perspicuity.

Further, Schrödinger felt his theory was more physically intuitive, thus providing a better foundation for solving and understanding physical problems, whereas Heisenberg's version was merely a picture of abstractions and complicated mathematics. In a letter to a friend he writes, "All philosophizing about 'principle observability' only glosses over our inability to guess the right pictures."

Indeed, Schrödinger felt that the practical and intellectual development of quantum mechanics was best served by his nearly visualizable wave mechanics approach.

> To me it seems extraordinarily difficult to tackle problems of [quantum mechanics], as long as we feel obliged on epistemological grounds to repress intuition in atomic dynamics, and to operate only with such abstract ideas as transition probabilities, energy levels, etc.

Heisenberg believed his theory captured the true essence of quantum mechanics, describing it as the "true theory of a discontinuum" and expressed similar resentment towards Schrödinger's wave mechanics. In a letter to his friend Wolfgang Pauli (1900–1958), Heisenberg writes:

> The more I think about the physical portion of the Schrödinger theory, the more repulsive I find it. … What Schrödinger writes about the visualizability of his theory "is probably not quite right," in other words it's crap.

And just when it seemed these two versions of quantum mechanics couldn't be more different, Schrödinger wrote a paper, *On the Relation of the Heisenberg-Born-Jordan Quantum Mechanics to Mine*, showing that – at least from a mathematical standpoint – they are the same.

For physicists this was great news, since it meant that they could use either approach to solve physical problems. In other words, most physicists were relieved they could use Schrödinger's version since the mathematics was much less cumbersome than the matrices of Heisenberg's theory and already familiar to them from physical problems they'd previously solved. Nonetheless, the physical interpretation of both theories was still up for grabs.

The Physical Consequences of Quantum Mechanics

The biggest question still plaguing Schrödinger's wave equation was the role of the wavefunction. Sure, mathematically it's clear: it's the solution to Schrödinger's wave equation and the "all-powerful function" as a result. However, physically it was still a big mystery to everyone, including Schrödinger himself.

Originally he had interpreted it as being connected to "some vibration process in the atom." Later, for a system of electrons, Schrödinger became more exacting and interpreted the absolute square of the wavefunction (the absolute value of the wavefunction multiplied by itself) as a type of "weight-function" related to the charge density (or "density of electricity," as he put it) at a particular region in space. Therefore, for a single electron, Schrödinger actually envisioned it to be spread out over the entire space. In other words, he literally imagined the electron not as a particle at a particular point in space, but as a wave spread throughout it.

Schrödinger wasn't the only one to be thinking about the physical meaning of the wavefunction. Several people were beginning to conclude that the wavefunction was really associated with a kind of *quantum probability* very different from the probability of classical mechanics. Among them were Paul Dirac (1902–1984), Eugene Wigner (1902–1995), and, most notably, Max Born. Max Born's work clearly defined the physical meaning of the wavefunction and the nature of quantum probability. He says, "The motion of particles follows probability laws …."

In other words, the motion of quantum particles, such as electrons, isn't governed by deterministic equations like classical particles (or objects). As a result, unlike a classical particle, a quantum particle doesn't move along a well-defined physical path with well-defined values for its key properties, such as position, momentum, energy, and the like, at every instant in time. Rather, according to Born, these physical quantities (and many others) are determined entirely by a quantum probability, which is proportional to the absolute square of the wavefunction. As it was for Schrödinger, so was it for Born that the absolute square of the wavefunction was the secret to revealing the true physical meaning of the wavefunction. However, his perception of it was altogether different.

Born also notes that "the probability itself propagates according to the law of causality." So, although the motion of a quantum particle isn't deterministic, the quantum probability governing the final outcome is, and it's given by Schrödinger's wave equation (since this determines the wavefunction, and therefore the absolute square of it). This is somewhat reminiscent of when we talked about the Boltzmann probability (in Part II). Recall that the Boltzmann probability gives the probability of a certain microstate occurring from the many that are available to a system of particles. However, there's a crucial difference.

The Boltzmann probability was a mathematical convenience for handling a system containing an extremely large number of particles. For such a system, it's simply impossible to use Newton's equation to determine the physical path of each and every particle. That doesn't mean that their paths and the respective positions, momentum, energy, and the like don't exist. Surely, in classical mechanics, they do exist. It just means that solving this mathematical problem is unwieldy. Thus, we appeal to using the Boltzmann probability, which greatly simplifies the original problem by allowing us to calculate average quantities of the collective system of particles.

It's a totally different scenario with the probability associated with quantum mechanics. Here, the quantum probability isn't merely a way of simplifying some complicated mathematics into a more tractable problem. In the quantum world, the probabilistic nature *is* the physical reality. Therefore, the *only* thing one can know about a quantum particle is the probability that you will find it in a given quantum (micro) state. So whereas in classical mechanics one talks of a particle evolving, according to Newton's equation, along a given path, in quantum mechanics it's the probability of the particle that evolves, according to Schrödinger's equation, leading it from one quantum state to another.

Born's interpretation reconciled a major challenge of Schrödinger's wave equation. Now, it was clear how the continuous wavefunction could give rise to the discontinuous energy states available to an electron: the wavefunction moves in a probability space – not a physical space – that "guides" the electron from one quantum state to another. So, the probabilistic nature of quantum, which Heisenberg had explicitly included in his theory, is also present in Schrödinger's theory. The electron jumping in Bohr's atom had also been reaffirmed by Born's interpretation; now these electronic transitions had been endowed with a mathematical probability of their occurrence. However, the electronic orbits (a purely classical notion), which had been falling out of favor for some time, were finally gone once and for all. For Schrödinger this was all too much.

One would have thought that Schrödinger would have been a big fan of Born's statistical interpretation. After all, this was the guy who, in his 1922 inaugural address at the University of Zürich, said, "It is quite possible that the laws of nature without exception have a statistical character." Evidently, quantum probability was not what he was thinking when he made this statement:

> But today I no longer like to assume with Born that an individual process of this kind is "absolutely random," i.e., completely undetermined. I no longer believe today that this conception (which I championed so enthusiastically four years ago) accomplishes much.

Schrödinger wasn't the only one to have trouble with quantum probability; Einstein did as well.

After leading the way in quantum theory for almost twenty years and being the first to introduce transition probabilities into it (in 1916), Einstein had had enough. In 1917, he noted that according to quantum theory, the direction of momentum transfer in the spontaneous emission of a photon from an atom was apparently governed by "chance," and it troubled him ever since. By 1926, he had become completely unforgiving of quantum probability, and in response to a letter Born had written to him, Einstein wrote:

> Quantum mechanics is very impressive. But an inner voice tells me that it is not yet the real thing. The theory produces a good deal but hardly brings us closer to the secret of the Old One. I am at all events convinced that *He* does not play dice.

Einstein's denouncement came as a "hard blow" to Born. It was official: Einstein was turning his back on quantum mechanics and so it would be the rest of his life. In 1944, he would reaffirm his original statement to Born: "The great initial success of the quantum theory cannot convert me to believe in that fundamental game of dice."

While Born and Einstein were disagreeing on quantum probability, Bohr and Schrödinger were disagreeing on – well, pretty much everything.

In September 1926, Bohr invited Schrödinger to Copenhagen to lecture on and discuss wave mechanics in more detail. No sooner had Schrödinger stepped off the train than Bohr began debating the physical interpretations of quantum mechanics. Over the next several days, Bohr became even more relentless with discussions, beginning early in the morning and continuing late into the night. Further, to eliminate the possibility of any distractions, Bohr arranged for Schrödinger to stay at his home. The most detailed account of this visit comes from Heisenberg, who at that time was Bohr's assistant at the institute. He recalled, "although Bohr as a rule was especially kind and considerate in relations with people, he appeared to me now like a relentless fanatic ...."

Schrödinger interpreted his wave equation literally, viewing a quantum object, like an electron, as a wave rather than a particle. For him, the wavefunction was really a matter wave, describing where various portions of an electron were actually scattered throughout space. Also, he insisted upon the possibility of some sort of visualizable construct of the inner workings of the atom.

In total opposition to this stood Bohr. He was in agreement with Born that the wavefunction was describing a quantum probability and not some sort of real physical wave. During their discussion, Bohr favored the particle concept for quantum objects, but later he would require *both* particle and wave concepts for a complete description – a full realization of Einstein's "fusion concept" of 1909. However, their biggest point of contention was those jumping electrons in Bohr's atomic model.

As far as Schrödinger was concerned, the whole idea of electrons hopping from one stable, or *stationary*, discrete quantum state to another was ridiculous: "You surely must understand, Bohr, that the whole idea of quantum jumps necessarily leads to nonsense. … In other words, the whole idea of quantum jumps is sheer fantasy."

Who could blame him? Bohr's atomic model offered no "real" explanation as to why electrons didn't fall into the nucleus as classical physics would have it. It simply imposed the concept of the stationary quantum state, which was the "magic orbit" that freed the electron from this tragic fate. And if this wasn't disturbing enough, one had to further accept that an electron could jump between these stationary quantum states by simply absorbing or emitting a photon without any regard for what's actually governing this process.

Bohr acknowledged that these were difficult concepts to accept but felt that Schrödinger's biggest problem with it all came back to his need for a working visual:

> it does not prove that there are no quantum jumps. It only proves that we cannot imagine them, that the representational concepts with which we describe events in daily life and experiments in classical physics are inadequate when it comes to describing quantum jumps. Nor should we be surprised to find it so, seeing that the processes involved are not the objects of direct experience.

Schrödinger continued to insist that removing the discrete quantum states by adopting his wave picture of an electron would resolve the "quantum weirdness." Bohr pointed out that both Planck's Radiation Law and Einstein's work on the interaction of light and matter required discrete quantum states. Surely, was Schrödinger going to tear down the very foundation of quantum mechanics? Moreover, Bohr contended that experiments had already confirmed this discrete nature of the atom in a variety of ways. To this a frustrated Schrödinger replied: "If all this damned quantum jumping were really here to stay, I should be sorry I ever got involved with quantum theory."

After a few days, Schrödinger fell ill and stayed in bed with a feverish cold; perhaps Bohr had finally worn him down. Mrs. Bohr took care of him, bringing him tea and cake by his bedside, all the while Bohr sat on the edge of the bed, continuing to argue, "But surely Schrödinger, you must see …."

But he couldn't see. Although the discussions had a huge impact on both of them, in the end no resolution could be found. Nonetheless, Bohr and Schrödinger remained friends.

Once Schrödinger left, the discussions continued, now between Bohr and Heisenberg. As Heisenberg recalled, "During the next few months [after Schrödinger's visit] the physical interpretation of quantum mechanics was the central theme of all conversations between Bohr and myself."

Just as it was between Bohr and Schrödinger, the conversations between Bohr and Heisenberg were just as tense, although now they spanned over months rather than a few days. By February 1927, they had had enough of each other, prompting Bohr to withdraw to a skiing trip in Norway and leaving Heisenberg finally alone with his own thoughts, much to his relief. Within a few days, Heisenberg was already onto something.

Recall that using Schrödinger's formulation, Born had concluded that a quantum particle didn't follow a deterministic path, as classical mechanics would have it. Rather, the quantum state it was in (and would be in later) was completely governed by an inherent quantum probability. Heisenberg wondered: What if we did actually try to measure the path of an electron, or more simply the position and momentum of the electron at the same instance in time? In an ingenious thought experiment, Heisenberg imagined a microscope capable of such a task.

Now, a microscope works because a photon (from a light source) has bounced (reflected) off an object (like an electron) and ended up passing through the lens where one's eye or some other detector sees it. From classical wave optics, it was already known that the uncertainty (resolving power) in the position of the object is *directly* related to the wavelength of the photon used. So, if you want better resolution of the object's position, you need to use a photon of smaller wavelength. In other words, if you want a more precise measurement, you need to use a "ruler" with more precise, or smaller, tick marks. Moreover, the uncertainty in the position was also known to by *inversely* related to the diameter of the lens used to collect those reflected photons we mentioned. Once again, this is all from classical wave optics – nothing quantum going on just yet, other than we're calling light a photon. Now that we have a handle on determining the electron's position, let's consider the momentum.

When the photon bounces off the electron, they exchange momentum in the collision. The momentum lost by the photon will equal that gained by the electron; the momentum is conserved. To be sure, if we can determine the change in the photon's momentum, we'll then be able to determine the momentum of the electron *before* the collision,[212] which will give us both the electron's position and momentum at the same time as desired. However, there's a bit of an issue. We don't actually know the direction of the incoming photon after it has bounced off the electron. All we know for sure is that it was in the "range of directions" that resulted in its passing through the lens, thus allowing us to determine the electron position (within the uncertainty noted before). Now, this range of directions can be made narrower in order to minimize the uncertainty in the photon's direction of momentum. All we have to do is make the lens diameter on the microscope smaller.

---

[212] The electron's momentum after the collision can be determined using a detector. The photon's momentum before the collision can be established by using a light source, like laser light, where both the magnitude and direction of the photon momentum is known. This leaves the determination of the photon's momentum after the collision in order to solve for the initial electron momentum.

Oh, but wait – the uncertainty in the position is inversely related to the lens diameter, and a smaller diameter will result in a larger uncertainty in position. We could use a photon of a smaller wavelength to fix this issue. Unfortunately, it turns out this will increase the uncertainty in momentum, thanks to – you guessed it – the quantum nature of the photon's momentum. Heisenberg was able to use his matrix mechanics approach to show that certain pairs of properties (e.g., position and momentum in the same direction) can't be determined to arbitrary precision. Specifically, he found that the product of their uncertainties couldn't be smaller than Planck's constant.[213] As a result, it means that as we make our knowledge of one of the properties more exact, we end up making our knowledge of the complementary one less so. So, we either know one property almost exactly and therefore know nothing about the other one, or we compromise and know a little bit about both properties.

This has nothing to do with our ability (or lack thereof) to measure them. Rather (going back to our example with the microscope), it means that a quantum particle such as an electron just doesn't have an exact position and momentum at a given instance in time; these exist for it only in a *fuzzy*, not well-defined, manner. Realize that whereas Born used quantum probability to eliminate determinism in quantum mechanics, *Heisenberg's Uncertainty Principle* does so without ever invoking any sort of probabilistic notions. Therefore, they serve as two independent strikes against causality in quantum. Perhaps the best example of quantum probability and wave-particle duality is illustrated by the *double-slit experiment*.

The Double-Slit Experiment

Imagine an "electron gun" that fires electrons towards a wall with two holes (or slits) that are at equal distance (*D*) away from it, and equal distance

---

[213] Today, we know that it's actually Planck's constant divided by 4π.

(D') away from the center of the wall (Figure 16.1). The electron gun is mounted on a turret, which moves back and forth from side to side, much like an oscillating fan. Given this motion, it's clear that we are not aiming the electrons at the holes; rather, they are simply being fired very much in a random fashion. The holes themselves are the same size and just big enough to let an electron through.

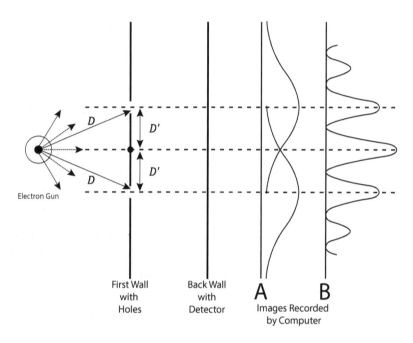

Figure 16.1. An "electron gun" fires at random upon a wall containing two holes that are at equal distance (D) away from it, and equal distance (D') away from the center of the wall. The electrons are stopped at the back wall, where a detector records their positions and sends this information to a computer. Image A is the distribution we get when we place detectors beside each hole to observe a passing electron. Here, we see no interference pattern, and in fact, we get the results we anticipated for an electron acting strictly as a particle, in which case it simply passes through one hole or the other. However, Image B is the distribution we get when there are no detectors present. Here, we do see an interference pattern as electrons pass through the holes.

As the electrons head towards the holes, some of them will pass through and some of them won't. The passing electrons continue on their way until they end up hitting yet another wall located much farther down[214] that acts as a backstop. At this back wall, the final position of each electron is recorded by a detector, which then sends this information to a computer for further processing.

As we continue firing more and more electrons (we want to get good statistics), more and more electrons pass through and hit the back wall. From the buildup of all the many electron positions, the computer is able to create a pattern, or distribution. If our statistics are good enough, then from this distribution we learn the probability of finding an electron at a given position on the back wall when fired randomly at the two holes. So, what does the distribution look like?

Before we get to this, let's take a moment to try and anticipate the results. Clearly, if an electron acts strictly as a particle, we would reasonably expect that it passes through one hole or the other. Moreover, an electron passing through a hole will either "bump" off of the side or edge, or will pass straight through totally unscathed. If it passes straight through, we'll find it directly behind the hole – at the "center," so to speak – when it hits the back wall, whereas if it gets bumped, we'll find it hits some distance farther away on either side of center. With all this in mind, we anticipate the distribution for a given hole to be such that the maximum number of hits occurs directly at the center, while farther away from there the number of hits steadily decreases. Lastly, the distribution will look the same on both sides of center. In other words, it will be *symmetric*.

OK, we have a pretty clear picture of what we'll see. But we perform the experiment and find that the resulting distribution on the computer screen is nothing like what we imagined. Instead, we find a distribution with the maximum located between the two holes – it's not even located at the center of either hole! The distribution is still symmetric on each side of this maximum (so at least there's that), but we don't see the steady decrease in the number of hits that we had envisioned as we move away. Instead, on either side we find peaks where the number of hits is high, and then from

---

[214] The wall is located at a distance much greater than the distance between the centers of the two holes.

these peaks there's a steady drop-off all the way to zero, where not a single electron shows up. What happened?

Well, in our foresight we assumed that an electron behaves as a particle, but we really should have known better since all quantum particles exhibit wave-particle duality. In short, the distribution formed by the collection of the many electron positions is showing an *interference* pattern. Earlier, we briefly talked about how interference can occur between waves. Evidently, there must be waves associated with our electrons that are causing this interference pattern. What are they? Recall that the position of each electron at the back wall will be determined by the quantum probability as we discussed earlier. In turn, the quantum probability is given by (the absolute square of the) wavefunction; it looks like we've found the "wave" causing the interference.

Let's try to understand this in more detail. Instead of shooting many electrons all at once towards the holes, let's just shoot one electron at a time. Initially, we notice that shortly after firing off an electron it arrives at the back wall and its position is detected. So far, so good. However, as we continue to shoot individual electrons at the holes, we noticed something quite peculiar. Eventually, we end up with the same interference pattern that we saw before when we were firing many electrons. In other words, it doesn't matter if we fire several electrons at once or one at a time; the same interference pattern appears! This means that a single electron – as it encounters the two holes – ends up interfering with itself.

This seems so odd that we decide to perform one last experiment to get to the bottom of things. Beside each hole we place a detector that will record an electron passing by it. Surely, this will shed some light on the strange results we are getting. Once again we shoot one electron at a time towards the holes, over and over, until we can clearly see the distribution on the computer screen. This time we find that the interference pattern has totally disappeared, and instead we are left with the distribution of electron positions that we had anticipated in the first place! In other words, when we're not looking at the holes (with our detectors) a single electron incurs interference, but when we do look, we find that the electron passes through either one hole or the other, and the interference pattern completely disappears.

These experiments illustrate the very essence of quantum mechanics. We see an electron acts as a particle when it hits the back wall and is detected by the detector as a localized entity, but somewhere in between there's interference due to its wave nature and its "interaction" with both holes at the same time. This wave nature is intimately tied to the quantum probability of finding the electron at a particular position on the back wall, which ultimately leads to the distribution of hits we see. If we attempt to determine exactly where an electron will end up on the back wall by trying to see which of the holes it passes through, the whole thing falls apart and the interference disappears altogether.

Although we chose to do our experiment with electrons, all quantum particles show this type of weird behavior. If all this seems like more science fiction than actual science to you, you're not alone. The physical consequences of quantum mechanics are, simply put, plain strange by comparison to our everyday experience.

A Game of Chance

Quantum theory and its successor, quantum mechanics, rattled the very core of our understanding of the physical world we live in. Energy, light, atoms, and matter – all of these major players came under heavy scrutiny. Even the very concepts that we use to describe some of these, like "wave" and "particle," were pushed to their breaking points, forcing us to now accept the existence of wave-particle duality for all quantum entities (electrons, light, and the like). And as if all that weren't enough, determinism itself, which had always been a part of classical physics, now had to be abandoned for uncertainty and an overarching quantum probability. These latter notions often cause the greatest confusion.

Heisenberg's Uncertainty Principle defines a rigorous physical restriction (imposed by nature) on how much we can know about certain pairs of variables, like an electron's position and momentum along a given direction. In other words, having better measuring devices will never reconcile this inherent uncertainty or deepen our knowledge. It also means that a quantum particle, such as an electron, simply doesn't have a well-defined trajectory it moves along. Rather, it "moves" between quantum states according to the quantum probability, which is related to Schrödinger's wavefunction. Bohr's atom with its "jumping electrons" – for all that it lacks – illustrates this quite well, much to Schrödinger's dismay.

While the quantum probability is reminiscent of the classical probabilities of Maxwell and Boltzmann, nothing could be further from the truth. For while the latter are self-imposed to ease the burden of complicated mathematics, they still preserve the underlying determinism so dear to classical physics. In contrast, quantum probability is an unequivocal affirmation by nature against determinism altogether. Indeed, reckless abandon of the well-oiled "world machine" for nothing short of a "game of chance" undoubtedly poses the greatest challenge to one's sensibilities. Nonetheless, by all current accounts quantum mechanics, with all its "weirdness" and probabilistic overtones, has endured the test of time.

Epilogue
From Here to There

Energy, entropy, atoms, and quantum mechanics form the major foundation upon which much of science is built today. Indeed, it's why I chose them as the major topics of the book. My overall goal was for you to know both the actual science and its amazing backstory. In this telling, we discussed a rich history and learned about the actual scientists who dealt with both personal and professional struggles to bring a sense of clarity to many important questions about the universe we all live in. In so doing, each of them built upon the work of others and harnessed their passion for knowing "something absolute" (as Planck said). It's in this way that science as a whole has progressed and will continue to move forward. We discussed the major contributors, but there were many others, over a span of many years, who added something, both big and small, to our overall understanding. I only regret that I couldn't detail them all here, but I am grateful for all their efforts and sacrifices. The areas of science I am most passionate about, chemistry and physics, continue to make great strides. Let me point you towards just a few.

In 1917, Einstein had already expressed an "uneasiness" with quantum mechanics and its inherent probabilistic nature. Nonetheless, he continued to make substantial contributions, with his last major effort being his three papers on the quantum ideal gas in 1924–1925. During the last thirty years of Einstein's life, his scientific activities focused on finding a *unified field theory*. It's reasonable to suspect that this was already on his mind at that turning point in 1917, and by the 1920s it became his main focus.

First and foremost, he was looking to unify gravity (as described by his very own general relativity) and electromagnetism (as described by Maxwell's equations). Moreover, he wanted the particles of physics to emerge as special solutions from the resulting equations of the theory. Last but not least, Einstein demanded that such a theory be completely causal. In other words, the new theory would explain (in addition to other things) all that quantum mechanics had so successfully been able to explain, just without the need for quantum probability.

On April 18, 1955, Einstein died at Princeton Hospital. Presumably working up until those last moments, Einstein, nonetheless, had failed to find the unified theory he so desired. However, attempts at such unification remain an active area of research pursued by physicists today, most notably in the way of *string theory*.[215]

If quantum mechanics with its quantum probability isn't strange enough for you, wait for this: a literal physical interpretation of the mathematics of Schrödinger's equation allows one to conceive of a quantum system as being in two or more quantum states – a superposition thereof – at the same time. This is, in the very least, jarring to our human sensibilities, as the world we experience doesn't seem to reveal such a crazy mixture of reality.

The standard thinking in quantum mechanics is that a given outcome occurs because the wavefunction "collapses" into a well-defined quantum state upon observation. Recall that when we discussed the double-slit experiment, we placed a detector beside each hole, hoping to gain insight into the puzzling interference behavior of the electrons as they passed through. Rather, all we saw was a well-defined quantum state whereby a single electron passed through either one hole or the other, and the interference pattern had now completely disappeared. The collapse of the wavefunction is part of the *Copenhagen Interpretation* of quantum mechanics and has remained the standard for some time.

However, Hugh Everett III (1930–1982) had a very different idea. Instead of artificially imposing the collapse of the wavefunction into a single quantum state upon observation, Everett opted for allowing all possible outcomes instead. The catch is that each outcome would take place in different *parallel universe*. So while we see a given outcome in *our* universe,

---

[215] For the reader interested in knowing more about this area of research, please have a look at Brian Greene's books.

the other outcomes also take place to would-be observers like us, or perhaps even other version of us, but in *other* universes. The concept of such a *multiverse* isn't unique to quantum mechanics. It pops up in the fields of relativity, cosmology, and others.[216]

In our everyday lives we have come to believe certain things will never occur: a broken glass won't reassemble itself; the creamer in one's coffee won't become unmixed; and the air on one side of the room won't completely move to the other side, leaving us breathless (thank goodness!). As we learned, such processes (or transitions) don't occur because irreversibility doesn't allow it. Moreover, this irreversibility is intimately tied to the universe's natural tendency to increase its entropy, something not at all described by the first law, but rather forming the underpinnings of the second law. Upon reflecting on, for example, the broken glass, one would surely conclude that at some time in the *past* it was unbroken. In other words, an *arrow of time*[217] has been drawn, whereby an increase in entropy is the signpost of the present or future, but not the past.

Boltzmann taught us that the favored macrostate of a system is the one with the most microstates, and consequentially the one with highest entropy (for the system's current set of conditions, e.g., temperature, pressure, etc.). Indeed, there are more microstates (the microscopic arrangement of atoms making up the glass) for a glass to be broken, whereas there is really only one way for it to be unbroken (to be sure, gluing it back together doesn't make it unbroken again). And while Boltzmann's explanation brings a certain amount of satisfaction, something fundamental is missing: What is the underlying mechanism that really drives the universe (or a system) towards increasing entropy in the first place? Honestly, we don't really know yet.

I could keep this going, but I leave you here with these, the many things we discussed already, and my hope is that it allows you to find a bit of passion of your own for science.

---

[216] I refer you to Brian Greene's *The Hidden Reality: Parallel Universes and the Deep Laws of the Cosmos.*

[217] I recommend *From Eternity to Here: The Quest for the Ultimate Theory of Time* by Sean Carroll.

Bibliography

Aitchison, Ian J. R., David A. MacManus, and Thomas M. Snyder. "Understanding Heisenberg's 'Magical' Paper of July 1925: A New Look at the Calculational Details." *American Journal of Physics* 72 (2004): 1370–1379.

Antognazza, Maria R. *Leibniz: An Intellectual Biography.* New York: Cambridge University Press, 2008.

Badino, Massimiliano. "Probability and Statistics in Boltzmann's Early Papers on Kinetic Theory." *PhilSci-Archive: An Archive for Preprints in Philosophy of Science.* http://philsci-archive.pitt.edu/2276/ (accessed 2016).

———. "The Odd Couple: Boltzmann, Planck and the Application of Statistics to Physics (1900–1913)." *Annalen Der Physik* 18 (2009): 81–101.

Banach, David. "Plato's Theory of Forms." *David Banach Homepage (Philosophy 105).* http://www.anselm.edu/homepage/dbanach/platform.htm (accessed 2016).

———. "Some Main Points of Aristotle's Thought." *David Banach Homepage (Philosophy 105).* http://www.anselm.edu/homepage/dbanach/arist.htm (accessed 2016).

Boas, Marie. "The Establishment of the Mechanical Philosophy." *Osiris* 10 (1952): 412–541.

Bose, Satyendra N. "Planck's Law and the Light Quantum Hypothesis (Translated by O. Theimer and Budh Ram)." *American Journal of Physics* 44 (1976): 1056–1057.

Boyle, Robert. *The Sceptical Chymist.* Mineola: Dover Publications, 2003.

Brush, Stephen G. "How Ideas Became Knowledge: The Light-Quantum Hypothesis 1905–1935." *Historical Studies in the Physical and Biological Sciences* 37 (2007): 205–246.

———. "The Development of the Kinetic Theory of Gases." *Archive for History of Exact Sciences* 12 (1974): 1–88.

———. *The Kinetic Theory of Gases: An Anthology of Classic Papers with Historical Commentary (History of Modern Physical Sciences, Vol. 1).* London: Imperial College Press, 2003.

Burnham, Douglas. "Gottfried Leibniz: Metaphysics." *The Internet Encyclopedia of Philosophy.* http://www.iep.utm.edu/leib-met/ (accessed 2016).

Campbell, Gordon. "Empedocles." *The Internet Encyclopedia of Philosophy.* http://www.iep.utm.edu/empedocl/ (accessed 2016).

Cassidy, David C. *Uncertainty: The Life and Science of Werner Heisenberg.* New York: W. H. Freeman, 1992.

Cercignani, Carlo. *Ludwig Boltzmann: The Man Who Trusted Atoms.* New York: Oxford University Press, 1998.

Clausius, Rudolf. *The Mechanical Theory of Heat: With Its Applications to the Steam-Engine and to the Physical Properties of Bodies.* Edited by Thomas A. Hirst. J. Van Voorst, 1867.

Cohen, S. Marc. "Aristotle's Metaphysics." *The Stanford Encyclopedia of Philosophy (Summer 2014 Edition).* Edited by Edward N. Zalta. http://plato.stanford.edu/archives/sum2014/entries/aristotle-metaphysics/ (accessed 2016).

Cottingham, John, ed. *The Cambridge Companion to Descartes.* Cambridge University Press, 1992.

Cropper, William H. *Great Physicists: The Life and Times of Leading Physicists from Galileo to Hawking.* New York: Oxford University Press, 2004.

———. *The Quantum Physicists: And an Introduction to Their Physics.* New York: Oxford University Press, 1970.

Darrigol, Olivier. "Continuities and Discontinuities in Planck's Akt der Verzweiflung." *Annalen der Physik* 9 (2000): 951–960.

———. *From c-Numbers to q-Numbers: The Classical Analogy in the History of Quantum Theory.* Berkeley: University of California Press, 1992.

———. "The Historians' Disagreements over the Meaning of Planck's Quantum." *Centaurus* 43 (2001): 219–239.

———. "The Origins of the Entropy Concept." Edited by J. Dalibard, B. Duplantier, and V. Rivasseau. *Poincaré seminar 2003: Bose-Einstein condensation, Entropy.* Basel: Birkhäuser, 2004, 101–118.

De Camp, L. Sprague. *The Ancient Engineers: Technology and Invention from the Earliest Times to the Renaissance.* New York: Sterling Publishing, 1990.

Debus, Allen G. "Paracelsus, Five Hundred Years: Three American Exhibits." *U.S. National Library of Medicine.* http://www.nlm.nih.gov/exhibition/paracelsus/index.html (accessed 2016).

Descartes, René. *Principles of Philosophy.* Translated by Valentine R. Miller and Reese P. Miller. Netherlands: Kluwer Academic Publishers, 1991.

Dirac, Paul A. M. "The Physical Interpretation of the Quantum Dynamics." *Proceedings of the Royal Society of London A* 113 (1927): 621–641.

Dobbs, Betty Jo T. *The Foundations of Newton's Alchemy.* Cambridge: Cambridge University Press, 1983.

Drake, Stillman. *Essays on Galileo and the History and Philosophy of Science.* Edited by Trevor H. Levere and Noel M. Swerdlow. Vol. 1. Toronto: University of Toronto Press, 1999.

———. *Essays on Galileo and the History and Philosophy of Science.* Edited by Trevor H. Levere and Noel M. Swerdlow. Vol. 2. Toronto: University of Toronto Press, 1999.

———. *Essays on Galileo and the History and Philosophy of Science.* Edited by Trevor H. Levere and Noel M. Swerdlow. Vol. 3. Toronto: University of Toronto Press, 2000.

———. *Galileo at Work: His Scientific Biography.* Mineola: Dover Publications, 1995.

———. *Galileo: A Very Short Introduction.* New York: Oxford University Press, 2001.

Eckert, Michael. "Max von Laue and the Discovery of X-Ray Diffraction in 1912." *Annalen der Physik* 524 (2012): A83–A85.

Einstein, Albert. *Letters on Wave Mechanics: Correspondence with H. A. Lorentz, Max Planck, and Erwin Schrödinger.* New York: Open Road, 2011.

———. *The Collected Papers of Albert Einstein, Volume 14: The Berlin Years: Writings & Correspondence, April 1923–May 1925 (English Translation Supplement)*. Edited by Diana K. Buchwald, József Illy, Zeʾev Rosenkranz, Tilman Sauer, and Osik Moses. Translated by Ann M. Hentschel and Jennifer N. James. Princeton: Princeton University Press, 2015.

———. *The Collected Papers of Albert Einstein, Volume 2: The Swiss Years: Writings, 1900–1909*. Translated by Anna Beck. Princeton: Princeton University Press, 1989.

———. *The Collected Papers of Albert Einstein, Volume 6: The Berlin Years: Writings, 1914–1917*. Translated by Alfred Engel. Princeton: Princeton University Press, 1997.

Fara, Patricia. *Newton: The Making of Genius*. New York: Columbia University Press, 2004.

Faraday, Michael. "On the Absolute Quantity of Electricity Associated with the Particles or Atoms of Matter." *Philosophical Transactions of the Royal Society of London* 124 (1834): 77–122.

Feynman, Richard P., Robert B. Leighton, and Matthew Sands. *The Feynman Lectures on Physics: Commemorative Issue, Three Volume Set*. Redwood City: Addison Wesley, 1989.

———. *The Character of Physical Law*. New York: Modern Library, 1994.

Fitzpatrick, Richard. "Heisenberg's Uncertainty Principle." *Quantum Mechanics*. http://farside.ph.utexas.edu/teaching/qmech/Quantum/node27.html (accessed 2016).

Fowler, Michael. "Evolution of the Atomic Concept and the Beginnings of Modern Chemistry." *Modern Physics*. http://galileo.phys.virginia.edu/classes/252/atoms.html (accessed 2016).

Galilei, Galileo. *Dialogue Concerning the Two Chief World Systems: Ptolemaic and Copernican*. Edited by Stillman Drake. New York: The Modern Library, 2001.

———. *Dialogues Concerning Two New Sciences*. Translated by Henry Crew and Alfonso de Salvio. New York: Prometheus Books, 1991.

————. *On Motion and Mechanics: Comprising De Motu (ca. 1590) and Le Meccaniche (ca. 1600).* Translated by I. E. Drabkin and Stillman Drake. Madison: The University of Wisconsin Press, 1960.

————. *The Essential Galileo.* Translated by Maurice A. Finocchiaro. Indianapolis: Hackett Publishing Company, Inc, 2008.

Galilei, Galileo, and Stillman Drake. *Two New Sciences/A History of Free Fall: Aristotle to Galileo.* Translated by Stillman Drake. Toronto: Wall & Emerson, Inc., 2000.

Gamow, George. *The Great Physicists from Galileo to Einstein.* Mineola: Dover Publications, 1988.

Gavin, Sean, and Stephen P. Karrer. "The Living Force." *Sean Gavin's Homepage.* http://rhig.physics.wayne.edu/~sean/Sean/Course_information_files/VisViva.pdf (accessed 2016).

Gearhart, Clayton A. "Planck, the Quantum, and the Historians." *Physics in Perspective* 4 (2002): 170–215.

Georgia State University. *HyperPhysics.* http://hyperphysics.phy-astr.gsu.edu/hbase/hph.html (accessed 2016).

Gindikin, Semyon G. *Tales of Physicists and Mathematicians.* Translated by Alan Shuchat. Boston: Birkhäuser, 1988.

Goldstein, Martin, and Inge F. Goldstein. *The Refrigerator and the Universe: Understanding the Laws of Energy.* Cambridge: Harvard University Press, 1995.

Graham, Daniel W. "Anaximenes." *The Internet Encyclopedia of Philosophy.* http://www.iep.utm.edu/anaximen/ (accessed 2016).

Gribbin, John. *Erwin Schrödinger and the Quantum Revolution.* Hoboken: Wiley, 2013.

Hankins, Thomas L. "Eighteenth-Century Attempts to Resolve the Vis Viva Controversy." *Isis* 56 (1965): 281–297.

Harman, Peter M. *Energy, Force and Matter: The Conceptual Development of Nineteenth-Century Physics.* Cambridge University Press, 1982.

Heilbron, John L. *Galileo.* Oxford: Oxford University Press, 2012.

Heisenberg, Werner. *Encounters with Einstein: And Other Essays on People, Places and Particles.* Princeton: Princeton Science Press, 1989.

———. *The Physical Principles of the Quantum Theory.* Translated by Carl Eckart and F.C. Hoyt. Mineola: Dover Publications, 1949.

Holmyard, Eric J. *Alchemy.* Mineola: Dover Publications, 1990.

Holton, Gerald, and Stephen G. Brush. *Physics, the Human Adventure: From Copernicus to Einstein and Beyond.* Rutgers University Press, 2001.

Iltis, Carolyn. "Bernoulli's Springs and their Repercussions in the Vis Viva Controversy." *Actes du XIIIe Congrès International d'Histoire des Sciences.* Moscow, 1974. 309–315.

———. "D'Alembert and the Vis Viva Contoversy." *Studies in History and Philosophy of Science* 1 (1970): 135–144.

———. "Leibniz and the Vis Viva Controversy." *Isis* 62 (1971): 21–35.

———. "The Controversy Over Living Force: Leibniz to D'Alembert (Doctoral Dissertation)." http://nature.berkeley.edu/departments/espm/env-hist/dissertation.html (accessed 2016).

Isler, Martin. *Sticks, Stones, & Shadows: Building the Egyptian Pyramids.* Oklahoma: University of Oklahoma Press, 2001.

Janiak, Andrew. "Newton's Philosophy." *The Stanford Encyclopedia of Philosophy (Summer 2014 Edition).* Edited by Edward N. Zalta. http://plato.stanford.edu/archives/sum2014/entries/newton-philosophy (accessed 2016).

Jones, Sheilla. *The Quantum Ten: A Story of Passion, Tragedy, Ambition, and Science.* New York: Oxford University Press, 2008.

Klein, Martin J. "Einstein's First Paper on Quanta." *The Natural Philosopher* 2 (1963): 59–86.

———. "Planck, Entropy, and Quanta, 1901–1906." *The Natural Philosopher* 1 (1963): 83–108.

———. "Thermodynamics in Einstein's Thought." *Science* 157 (1967): 509–516.

Kleppner, Daniel. "Rereading Einstein on Radiation." *Physics Today*, 2005: 30–33.

Kuhn, Thomas S. *Black-Body Theory and the Quantum Discontinuity, 1894–1912*. Chicago: University of Chicago Press, 1987.

———. *The Structure of Scientific Revolutions*. 4th ed. Chicago: University of Chicago Press, 2012.

Laidler, Keith J. *The World of Physical Chemistry*. Oxford: Oxford University Press, 1995.

Lau, Katherine I., and Kim Plofker. "The Cycloid Pendulum Clock of Christiaan Huygens." In *Hands on History: A Resource for Teaching Mathematics*, edited by Amy Shell-Gellasch, 145–152. Washington, DC: Mathematical Association of America, 2007.

Leibniz, Gottfried W. *Leibniz: Philosophical Essays (Hackett Classics)*. Translated by Roger Ariew and Daniel Garber. Indianapolis: Hackett Publishing Company, 1989.

Linden, Stanton J. *The Alchemy Reader: From Hermes Trismegistus to Isaac Newton*. Cambridge: Cambridge University Press, 2003.

Lindsay, Robert B. "The Concept of Energy and Its Early Historical Development." *Foundations of Physics* 1 (1971): 383–393.

Loschmidt, Johann. "On the Size of the Air Molecules." Edited by William W. Porterfield and Walter Kruse. *Journal of Chemical Education*, 1995: 870–875.

Müller, Ingo. *A History of Thermodynamics: The Doctrine of Energy and Entropy*. Berlin: Springer Berlin Heidelberg, 2007.

MacDougal, Douglas W. *Newton's Gravity: An Introductory Guide to the Mechanics of the Universe*. New York: Springer, 2012.

Mach, Ernst. *History and Root of the Principle of the Conservation of Energy*. Translated by Philip E. B. Jourdain. Chicago: Open Court Publishing Company, 1911.

Machamer, Peter. "Galileo Galilei." *The Stanford Encyclopedia of Philosophy (Winter 2014 Edition)*. Edited by Edward N. Zalta. http://plato.stanford.edu/archives/win2014/entries/galileo (accessed 2016).

Mahoney, Michael S. "Christiaan Huygens, The Measurement of Time and Longitude at Sea." In *Studies on Christiaan Huygens*, edited by H. J .M. Bos, J. S. Rudwick, H. A. M. Melders, and R. P. W. Visser, 234–270. Lisse: Swets & Zeitlinger, 1980.

Maxwell, James C. "Molecules." *Nature* VIII (1873): 437–441.

———. "The Theory of Molecules." *Popular Science Monthly* 4 (1874): 276–290.

McDonough, Jeffrey K. "Leibniz's Philosophy of Physics." *The Stanford Encyclopedia of Philosophy (Spring 2014 Edition)*. Edited by Edward N. Zalta. http://plato. stanford.edu/archives/spr2014/entries/leibniz-physics (accessed 2016).

Mehra, Jagdish, and Helmut Rechenberg. *The Historical Development of Quantum Theory, Vol. 4: Part 1: The Fundamental Equations of Quantum Mechanics 1925-1926; Part 2: The Reception of the New Quantum Mechanics 1925-1926*. New York: Springer-Verlag, 2001.

———. *The Historical Development of Quantum Theory, Vol. 5: Erwin Schrödinger and the Rise of Wave Mechanics. Part 2: The Creation of Wave Mechanics: Early Response and Applications 1925-1926*. New York: Springer-Verlag, 2001.

Meldrum, Andrew N. *Avogadro and Dalton: The Standing in Chemistry of Their Hypotheses*. University of Aberdeen, 1904.

Meyer, Joseph. "Roman Siege Machinery and the Siege of Masada." *2012 AHS Capstone Projects, Paper 14*, 2012.

Moore, Walter J. *Schrödinger: Life and Thought*. New York: Cambridge University Press, 1992.

Moran, Bruce T. *Distilling Knowledge: Alchemy, Chemistry, and the Scientific Revolution*. Cambridge: Harvard University Press, 2005.

Muir, Matthew Moncrieff Pattison. *A History of Chemical Theories and Laws*. New York: Wiley, 1907.

Myers, Richard L. *The Basics of Chemistry*. Westport: Greenwood Press, 2003.

Newburgh, Ronald, Joseph Peidle, and Wolfgang Rueckner. "Einstein, Perrin, and the Reality of Atoms: 1905 Revisited." *American Association of Physics Teachers* 74 (2006): 478–481.

Newman, William R. *Atoms and Alchemy: Chymistry and the Experimental Origins of the Scientific Revolution.* Chicago: University of Chicago Press, 2006.

Newman, William R., and Lawrence M. Principe. *Alchemy Tried in the Fire: Starkey, Boyle, and the Fate of Helmontian Chymistry.* Chicago: University of Chicago Press, 2002.

———. "Alchemy vs. Chemistry: The Etymological Origins of a Historiographic Mistake." *Early Science and Medicine* 3 (1998): 32–65.

Newton, Isaac. *The Principia.* Translated by Andrew Motte. New York: Prometheus Books, 1995.

O'Connor, John J., and Edmund F. Robertson. "Louis Victor Pierre Raymond duc de Broglie." *The MacTutor History of Mathematics Archive.* http://www-history.mcs.st-andrews.ac.uk/Biographies/Broglie.html (accessed 2016).

O'Keefe, Tim. "Epicurus." *The Internet Encyclopedia of Philosophy.* http://www.iep.utm.edu/epicur/ (accessed 2016).

O'Raifeartaigh, Lochlainn. *The Dawning of Gauge Theory.* Princeton: Princeton University Press, 1997.

Pais, Abraham. *The Genius of Science: A Portrait Gallery of Twentieth-Century Physicists.* New York: Oxford University Press, 2000.

———. *Inward Bound: Of Matter and Forces in the Physical World.* New York: Oxford University Press, 1988.

———. *Niels Bohr's Times: In Physics, Philosophy, and Polity.* New York: Oxford University Press, 1993.

———. *Subtle Is the Lord: The Science and the Life of Albert Einstein.* New York: Oxford University Press, 1982.

Patzia, Michael. "Anaxagoras." *The Internet Encyclopedia of Philosophy.* http://www.iep.utm.edu/anaxagor/ (accessed 2016).

Principe, Lawrence M. "Reflections on Newton's Alchemy in Light of the New Historiography of Alchemy." In *Newton and Newtonianism: New Studies,* edited by James E. Force and Sarah Hutton, 205–19. Dordrecht: Kluwer Academic Publishers, 2004.

Purrington, Robert D. *Physics in the Nineteenth Century.* New Brunswick: Rutgers University Press, 1997.

Robinson, Andrew. *Einstein: A Hundred Years of Relativity.* New York: Harry N. Abrams, Inc, 2005.

Ross, Sydney. "John Dalton." *Encyclopædia Britannica Online.* http://www. britannica.com/EBchecked/topic/150287/John-Dalton/217770/Atomic-theory (accessed 2016).

Rubenstein, Richard E. *Aristotle's Children: How Christians, Muslims, and Jews Rediscovered Ancient Wisdom and Illuminated the Middle Ages.* Orlando: Harcourt, 2004.

Russell, Bertrand. *The History of Western Philosophy.* New York: Simon and Schuster, 1945.

Ryckman, Thomas. *The Reign of Relativity: Philosophy in Physics 1915–1925.* New York: Oxford University Press, 2005.

Scerri, Eric R. *The Periodic Table: Its Story and Its Significance.* New York: Oxford University Press, 2007.

Schmaltz, Tad M. *Descartes on Causation.* New York: Oxford University Press, 2013.

Schrödinger, Erwin. *Collected Papers on Wave Mechanics.* Translated by J.F. Shearer and W.M. Deans. Providence: AMS Chelsea Publishing, 1982.

———. "On Einstein's Gas Theory (English Tranlsation by T. C. Dorlas)." *Teunis (Tony) C. Dorlas Homepage.* http://homepages.dias.ie/dorlas/Papers/ schrodinger_gas.pdf (accessed 2016).

Shapin, Steven. *The Scientific Revolution.* Chicago: University of Chicago Press, 1998.

Slowik, Edward. "Descartes' Physics." *The Stanford Encyclopedia of Philosophy (Summer 2014 Edition).* Edited by Edward N. Zalta. http://plato.stanford. edu/archives/sum2014/entries/descartes-physics/ (accessed 2016).

Smith, George E. "The Vis Viva Dispute: A Controversy at the Dawn of Dynamics." *Physics Today* 59 (2006): 31–36.

Sorell, Tom. *Descartes: A Very Short Introduction.* Oxford: Oxford University Press, 2000.

Stokes, Philip. *Philosophy: 100 Essential Thinkers.* New York: Enchanted Lion Books, 2005.

Straub, William O. "On the Failure of Weyl's 1918 Theory." *viXra.org e-print archive.* http://vixra.org/abs/1401.0168 (accessed 2016).

Terrall, Mary. "Vis Viva Revisited." *History of Science* 42 (2004): 189–209.

Uffink, Jos. "Boltzmann's Work in Statistical Physics." *The Stanford Encyclopedia of Philosophy (Fall 2014 Edition).* Edited by Edward N. Zalta. http://plato.stanford.edu/archives/fall2014/entries/statphys-Boltzmann/ (accessed 2016).

van der Waerden, Bartel L. *Sources of Quantum Mechanics.* Mineola: Dover Publications, 2007.

van Melsen, Andrew G. *From Atomos to Atom: The History of the Concept Atom.* Mineola: Dover Publications, 2004.

Westfall, Richard S. *Never at Rest: A Biography of Isaac Newton.* New York: Cambridge University Press, 1983.

Weyl, Hermann. *Mind and Nature: Selected Writings on Philosophy, Mathematics, and Physics.* Edited by Peter Pesic. Princeton: Princeton University Press, 2009.

Index

75645822R10199

Made in the USA
Middletown, DE
07 June 2018